MODELS OF PREVENTIVE MAINTENANCE

STUDIES IN MATHEMATICAL AND MANAGERIAL ECONOMICS

Editor

HENRI THEIL

VOLUME 23

NORTH-HOLLAND PUBLISHING COMPANY
AMSTERDAM – NEW YORK – OXFORD

MODELS OF PREVENTIVE MAINTENANCE

I. B. GERTSBAKH
Ben Gurion University of The Negev

1977

NORTH-HOLLAND PUBLISHING COMPANY
AMSTERDAM – NEW YORK – OXFORD

North-Holland ISBN series 07204 3300 2
North-Holland ISBN volume 07204 0465 7

Publishers:
NORTH-HOLLAND PUBLISHING COMPANY
AMSTERDAM – NEW YORK – OXFORD

Sole distributors for the U.S.A. and Canada:
ELSEVIER/NORTH-HOLLAND INC.
52 VANDERBILT AVENUE
NEW YORK, N.Y. 10017

Library of Congress Cataloging in Publication Data

Gertsbakh, Il'ia Borukhovich.
Models of preventive maintenance.

(Studies in mathematical and managerial economics; v. 23)

Translation of Modeli profilaktiki.
Bibliography: p. 249
Includes index.
1. Plant maintenance – Mathematical models. 2. Stochastic processes. 3. Reliability (Engineering)
I. Title. II. Series.
TS192.G4731 658.2'02'0184 76-7931
ISBN 0 7204 0465 7

Printed in The Netherlands

INTRODUCTION TO THE SERIES

This is a series of books concerned with the quantitative approach to problems in the social and administrative sciences. The studies are in particular in the overlapping areas of mathematical economics, econometrics, operational research, and management science. Also, the mathematical and statistical techniques which belong to the apparatus of modern social and administrative sciences have their place in this series. A well-balanced mixture of pure theory and practical applications is envisaged, which ought to be useful for universities and for research workers in business and government.

The Editor hopes that the volumes of this series, all of which relate to such a young and vigorous field of research activity, will contribute to the exchange of scientific information at a truly international level.

The Editor

PREFACE

Today the need for better organization of servicing of complex technical equipment has become most urgent. This explains the increasing interest in *reliability theory* and especially that part of the theory which deals with "optimal preventive maintenance".

Preventive maintenance, broadly speaking, is the total of all service functions aimed at maintaining and improving reliability performance characteristics and concerns itself with such activities as the replacement and renewals of elements, inspections, testing and checking of working parts during their operation.

There always exists a possibility for the repair team to vary both the number of times that preventive action is taken and the "depth" (thoroughness), that is, the degree of renewal, number of replaced elements, and so on. This permits the selection of the best preventive maintenance policy. Actual circumstances determine the most suitable criterion for optimal performance: in one case, minimal service cost, in another, top operational performance, etc.

This book is designed to provide mathematical descriptions of different types of preventive service situations. When a mathematical model of some existing phenomenon is constructed, it is necessary to consider only its most essential and significant features in order to provide the possibility of further formal mathematical treatment. Several mathematical models of preventive maintenance are presented in this book which describe situations in stochastically failing equipment. These models serve as tools to provide the optimal actions for preventive maintenance.

For the application of all models to practical situations, it is necessary to have more or less reliable statistical data concerning the lifetime distributions and behaviour of randomly changing parameters, etc. Collecting and processing of this statistical data forms a separate topic which is given only slight consideration in this book. It is assumed that the reader has a modest acquaintance with probability theory, mathematical statistics, differential and integral calculus, and is familiar with the basic ideas of reliability theory.

Chapter 1 gives some typical situations of preventive maintenance, all having the following common features: the behaviour of the object under consideration is a stochastic one; it is necessary to select some service rule for the preventive maintenance of this object; the choice of this rule must be subject to an appropriate criterion of optimality. Therefore, the problem of finding the optimal preventive maintenance policy is treated as a problem in finding the optimal control of some random process. This book is devoted to those models which can be treated within the framework of controlled semi-Markov processes. Section 3 of Chapter 1 contains the main facts of this theory. A knowledge of this section is a necessary prerequisite for understanding the contents of this book.

Each section in Chapters 2, 3 and 4 is devoted to a discussion of a specific probabilistic preventive maintenance model. The exposition of each section is almost independent, and only in rare cases are there references to preceding sections. This makes it possible for the reader to study the topic which is of interest to him immediately after his acquaintance with the contents of Chapter 1.

Chapter 2 is devoted to models which have the following characteristic features: the system can be classified into only two states – good and failure. The *a priori* data is the distribution function of the systems lifetime or of the failure-free operation time. A very typical model of this kind is the well-known age replacement where an operating unit is replaced by a new one if it fails or if its age reaches some prescribed value.

Chapter 3 deals with preventive maintenance when the system has, besides the good and failure states, several intermediate states which make it possible to judge "how close the system is approaching failure". A model of a system having a special prognostic parameter is introduced. The preventive action is carried out when a parameter which is stochastically connected with the failure enters a marginal domain.

For every model under consideration, a general description is given and the formal statement of the problem is described. Then usually follows an expression for the optimality criterion, and the algorithm for finding the optimal service rule is given. Every model is accompanied by a numerical example which demonstrates how the proposed algorithm works. Several most popular and useful models are supplied with tables containing data on optimal preventive maintenance rules.

Chapter 4 contains an exposition of two topics which are of interest for practical application of optimal preventive maintenance. The first one touches upon the problem of optimal grouping of the elements of a "complex" system according to their preventive repair periods in an hierarchically organized system. The second topic is a general approach to preventive maintenance of objects with multidimensional state description. A special scalarization procedure is proposed and methods for obtaining the best scalar parameter are proposed.

This book is not written in a purely theoretical style. Wherever it is not out of place, the exposition is accompanied by comments on the physical meaning of the results and the formal constructions are illustrated by heuristic arguments. Several statements from probability theory are cited without complete proofs; the necessary formal details can be found in the references. To simplify the reader's acquaintance with the contents of the book, some proofs have been given in the Appendix.

The author hopes that this book may be useful to applied mathematicians, management scientists and operations researchers who are interested in theoretical studies and in practical applications of reliability theory and, in general, in probabilistic methods of operations research. Presumably, this book may serve also as an introductory course in the theory of controlled Markov processes for operations research students.

I. B. Gertsbakh

CONTENTS

Chapter 1

INTRODUCTION TO THE THEORY OF CONTROLLED SEMI-MARKOV PROCESSES – BASIC CONCEPTS IN PREVENTIVE MAINTENANCE

1. Several Typical Examples of Preventive Maintenance

Two approaches predominate in reliability theory. The first consists of formulation of the problem as follows: which structural, technological and organizational measures will ensure the required reliability standard of the manufactured system when the system starts to work or is transmitted to its user. The choice of criteria which are most suitable for the system; elaboration of the system's structure to fit the reliability requirements; determination of the quality and operating conditions of its elements; reliability computation; life testing and marginal testing of the system and its parts, quality control of what is produced – all these are far from the full listing of those problems which must be solved at the stage of creation and designing of the system, before its use.

However, it is well known that the effectiveness of a working system depends not only on the "innate" properties built into it in the designing and production stages but on the quality of its operation, maintenance, repair, etc. If, for example, all maintenance actions are limited to emergency repairs made after the system has failed, then the operating characteristics of the system will probably be very low and it will not work efficiently. In addition to the first approach, there is a second "exploitational" approach. In its framework the problem is stated as follows: given the functioning system, we must elaborate measures to obtain the best possible operating characteristics. The pertinent circle of questions which must be solved in this connection is very wide. Included are: construction of the equipment used for inspecting and testing the parameters of the functioning system, working out methods of rapidly locating failures occurring during the operation, organizing parallel functioning of subsystems to provide standby reserve, problems of the optimal supply of spare parts, and finally, concerning all preventive maintenance measures – when to carry them out and of what do they consist.

Only a small part of the above-mentioned problems is examined in this book, specifically, the choice and substantiation of preventive maintenance actions which are best from the point of view of the systems' reliability or operational cost.

The exposition is based on the examination of several simplified schemes which we will designate as *models*. In reliability theory, as in any other theory, the objects under examination are not the real phenomena by themselves but those which remain after elimination of the nonimportant, secondary features of these phenomena and formalization of the most important. The patterns which are obtained as a result of such "extraction" and formalization are called *models* of phenomena. In this book we shall deal with models of systems which are adapted to the study of preventive maintenance.

The construction of a good-working formal model is not simple. On one hand, the model must be sufficiently complex to preserve by its construction all the essential characteristics of the phenomena under study. On the other hand, the purpose of the construction of the model is to achieve a quantitative study of the phenomena. That means that the model must be sufficiently simple to permit a formal mathematical description of the process under study. Now we will examine some typical situations arising in the process of servicing functioning systems. This will serve as an introduction to the more detailed study of models of preventive maintenance.

(A) Replacement of a Cutting Tool on the Automatic Production Line

In use, metal-cutting tools (cutters, milling-cutters, etc.) wear down and break. Because of the action of a large number of random factors (the quality of the instrument, its condition, the firmness of billet, etc.), the breakdown of the instrument (wear and tear beyond allowable norms, chipping, breakage) occurs after a random amount of time following the beginning of the operation. A breakdown occurring during the operation time leads to quite a large loss of working time, caused by the discovery and removal of the damaged part. During that time, the machine or the automatic production line is either idle or turning out defective products. Therefore, preventive replacement at a previously designated point of time is applied, requiring stoppage of the operation for a comparatively small time. The following question must be decided: how often must there be preventive replacement of tools in order to achieve maximal usage of the time during which the equipment must operate? It is difficult to answer that question immediately. If frequent replacements are prescribed, some time will be lost in making them. However, there will be almost no total breakdowns. On the contrary, a too long waiting period between planned replacements will cause significant losses of time because of breakdowns. It must be noted that the solution of the problem requires certain data concerning the reliability of the tools. Let us assume that the Reliability Department has had an opportunity to test a sample of the working tools and has obtained the lifetime distribution function $F(t)$. A slight formalization of the problem brings us to the following quite simple but practical important model of preventive maintenance.

Given the object (let us call it the element) with known distribution function (DF) of time-to-failure (often the word "lifetime" will be used). At the moment of failure, some sign of it becomes apparent and the failed element is replaced by a new one. This kind of repair we shall call an emergency repair (ER). Moreover, there is also a possibility of making preventive replacement or repair (PR) before we are aware of the failure. In fact, PR means the replacement of a *nonfailed* element. Both ER and PR effect a complete "renewal" of the element or, in other words, the replaced element at the beginning of its action has the same lifetime DF $F(t)$. During the ER, the element does not work for the period of time t_a; during the PR, for the period of time t_p. Preventive repairs are carried out at that instant when the total time of failure-free operation becomes equal to a multiple of a certain constant T (see Figure 1). This value T, called the

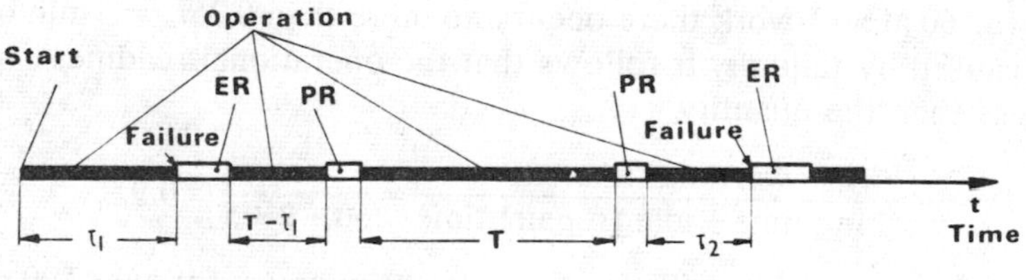

Figure 1. Scheme of tool replacement.

period of replacement, must be chosen so that the *efficiency* reaches its maximum. In this context, efficiency refers to the ratio of the time of failure-free operation to the sum of the time of failure-free operation plus the replacement (repair) time. This characteristic is called the coefficient of operational readiness, or simply, operational readiness (OR). We can show by a simple numerical example, that it is worthwhile to concern ourselves with the solution of this problem. For example, let our element have the lifetime DF shown in Figure 2; the mean lifetime is $t = 90$ min, $t_a = 20$ min, $t_p = 2$ min. Let us compare two cases: $T = 60$ min and $T = \infty$ (the PR is not done in general). In the first case, for every 60 min of working time, 2 min will be lost for performing PR and also some time for ER performed when failures occur between the PRs. It can be shown that

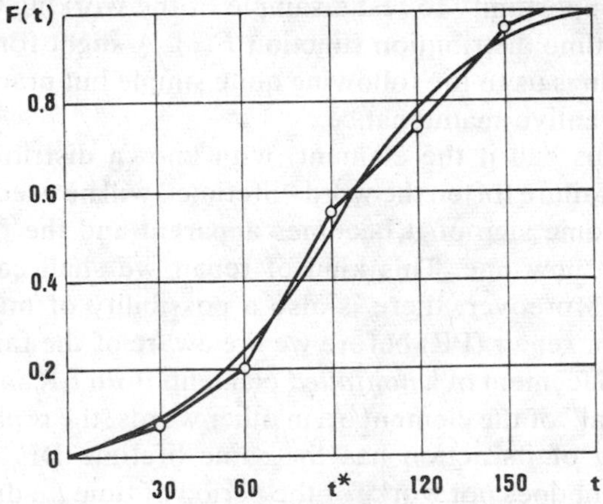

Figure 2. The DF of the failure-free operation time.

the average number of failures occurring over the interval T of operation time does not exceed the ratio $F(T)/(1 - F(T)) = 0.2/0.8 = 0.25$. Therefore, for 60 min of work there occurs no more than $0.25t_a = 5$ min of idle time caused by failures. It follows that the operational readiness will be not less than the quantity

$$k = \frac{\text{working time}}{\text{working time} + \text{idle (repair) time}} = \frac{60}{60+2+5} \cong 0.9.$$

If it will be assumed that $T = \infty$, then, on the average, for every 90 min of time-to-failure, 20 min will be lost for ERs. In that case we obtain a significantly smaller value $k = 90/(90 + 20) = 0.82$.

(B) Testing of "One-shot" Equipment

Several systems of a military nature are intended only for "one-shot" use. Until the moment when it is necessary to initiate the operation of the system, it is kept in reserve in a condition of partial readiness. As a result of inevitable aging of its elements, corrosion, etc., a malfunction may appear. The basic service demanded from the repair team is that they provide the maximal possible probability at a certain moment (not known beforehand) that the system is in good working order and fulfills its military task. This is insured by periodic inspections and testing of the state of the system, the values of its output characteristics, etc., and the resulting elimination of all discovered malfunctions or even failures. As to the choice of the inspection period, the problem resembles the previous one. It is obvious that too frequent checks are disadvantageous: the system may be needed right at the time when the inspection is being made. To completely exclude checking is also highly dangerous because the system may be found in a nonoperating (failure) state at the needed moment. In order to solve the problem of the proper time for optimal inspection, it is necessary to have at one's disposal data on the reliability of the system, element failure rates and time spent inspecting, testing and repairing. In Section 2 of Chapter 2 a formal model corresponding to this situation will be given.

(C) Periodic Testing of Output Characteristics of Radio and Electronic Equipment

One of the most frequently occurring malfunctions in radio and electronic equipment is the gradual drift of main output characteristics outside permissible levels. These may include the coefficient of amplification, parameters of electronic tubes, output power, etc. The "drift" is caused by a gradual, random process of aging or accumulation damages (see [4, Chapter 3] and [68, Chapter 4]). Special devices are used for periodic testing of values of the most important output characteristics. Assume, for example, that the equipment has only one important characteristic, denoted $\eta(t)$, and that the failure is defined as an event "$\eta(t) > M$" (see Figure 3). The testing is made periodically, at times $T, 2T, \ldots$. The maintenance rule is as follows: if the testing has discovered the state $\eta(t) < M_c$ (M_c is the "critical" level), nothing has to be done and the system is put into operation. If it happens that $M_c \leq \eta(t) \leq M$, preventive repair is made, as a result of which the value of the parameter $\eta(t)$ goes back to its initial level or near it. Every repair is associated with some cost c_0 and every failure occurring between checkups is considered as a breakdown of a normal operation and causes

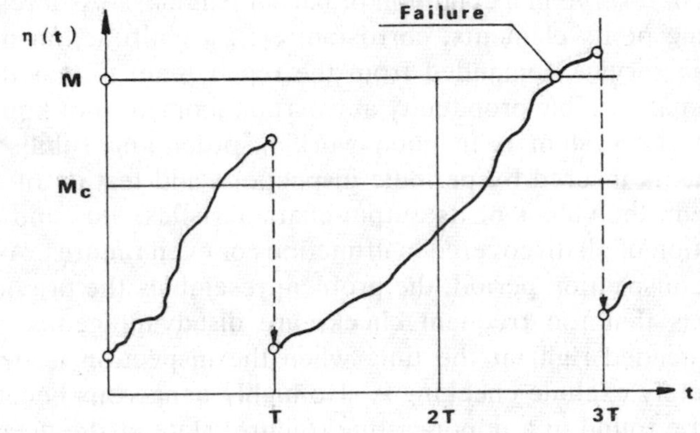

Figure 3. Preventive maintenance based on observations of the system's output parameter.

a payment c_1 (of course, $c_1 > c_0$). The value c_1 usually depends on the amount of time the system has spent in the inoperable state (up to the level M). The problem may be formulated as follows: find the optimal checking period T and the optimal "critical" level M_c which minimize the total average cost per unit operation time. In Section 7 of Chapter 3 a formal model corresponding to the above situation will be discussed.

(D) *Control of an Automatic Process of Production Based on the Current Quality Control Data*

Consider an automatic machine producing certain articles. During the operation under normal circumstances, the distribution of the size of the produced articles is characterized by certain parameters and a relatively small rate of defective pieces. When the machine is not performing normally, the rate of defective pieces increases significantly. To determine the exact moment of malfunction, samples of finished articles are taken periodically in order to carry out the necessary measurements. The examiner uses a predetermined rule for stopping the machine. For example: if a sample of 20 units has at least one defective unit, the process must be stopped for inspection and possibly for repairing the machine. Every such rule permits two kinds of errors: existing malfunction not being discovered, or on the contrary, a "false alarm", when the "defective" unit is discovered due to a measuring error. In both cases, losses occur – either because, although the equipment is working, it is producing defective articles or because it is shut down for an unnecessary inspection. In the above situation it is desirable to determine the optimal current quality control rule which includes the period of taking samples and a rule for deciding if the ongoing production process is in a good state or not. The "optimality" in this case means the maximal output over the given operation time.

This example is essentially more complicated than the previous ones. As a matter of fact, in the controlled process, the precision of the automatic machine is not observed directly. Available to us are only "indirect evidences" of the machine malfunctions – the result of testing the samples. This information only predicts the state of the machine to the degree that the larger the number of defective units, the higher the probability that the machine is out of order. Therefore, there is only a stochastic relationship between the information received and the actual state of the machine.

In Section 6 of Chapter 2, we shall propose a model dealing with this situation where the decision will be made on the basis of observing a series of consecutively produced defective articles.

Conclusions

(1) If we make a general abstraction from the specific contents of the examples given in Examples A–D, we can say that in all cases an object with nondeterministic behaviour was under consideration. Stochastic properties of that object were more or less known. In Examples A and B, only modest information was available: it was assumed that the DF of time-to-failure of the system or element was known. What goes on "inside" the system, how it approaches failure, remains unknown. In Example C, we were more informed about the processes occurring in the system because we had the opportunity to observe and therefore to study the behaviour of the system's main output characteristics $\eta(t)$. In Example D, the process $\eta(t)$ – the precision of the machine – is not observed directly but it is possible to observe another one – the number of defective articles stochastically connected with the main process.

(2) The personnel making repairs (we call them "the repair team") always have at their disposal a certain choice of possible repair actions including emergency and preventive measures. The latter include inspec-

tions, testing, preventive replacement of elements, etc. It is important that the repair team be able to select the parameters of the replacement policy: to choose the inspection period, the PR timing (Example A), to appoint the critical level for the controlled parameter (Example C), or to vary the size of the sample and the stopping rule (Example D). In other words, there is always the possibility of selecting a definite *service policy* to be carried out.

(3) Each of the actions mentioned in paragraph 2 is associated with some losses either in time or in money. When an automatic production line is stopped for PR, time is lost; when inspection of complex equipment is made, both time and money are forfeited.

(4) In all the above examples, the choice of the service policy has to be subject to certain optimality criterion. In Example A, it was necessary to maximize the operational readiness; in Example B, it was desirable to maximize the probability of fulfilling the military task of the equipment.

It is clear that in all the above situations we dealt, in fact, with some specific problems of *finding an optimal control policy for a random process*. We feel that determining the optimal preventive maintenance must be considered as a special kind of optimal control problem.

2. Terminology, Criteria of Optimality, and Classification of Preventive Maintenance Problems

2.1. Terminology

We shall distinguish only two categories of objects: systems and elements. A system is an object consisting of several devices or elements. It is assumed that for systems and elements there is a certain defined state – "failure" (where necessary a more detailed description of this state will be given). If a system or element has a special device which makes us aware of the failure, we shall refer to this as a "failure-signal" or "signalling system".

All the total repair actions directed at discovering the source of the failure and the subsequent repair will be called emergency repair (ER). The term preventive repair (PR) will be used to denote all kinds of repair actions directed at improvement of performance characteristics of the system undertaken in the absence of failure. Execution of PR means usually the replacement of deteriorated, aged but nonfailed elements by new ones or the improving of values of the system's output characteristics by tuning, adjusting, regulating, etc. If at the same time ER is being done for some elements and PR is being done for others, then this kind of repair action will be called emergency-and-preventive repair (EPR). The measures directed only at discovering the system's actual state will be called an inspection. Inspection gives information on whether or not the inspected element has failed, and on the values of system's output characteristics. For the sake of brevity we shall use the word "parameter" instead of "system's output", "performance characteristics", etc. It will be assumed in following chapters that PR and ER entirely renew the system; otherwise, a special explanation will be given.

All service measures can be divided into two groups – the preplanned and the nonplanned. The latter are usually caused by failures indicated by a failure-signal device. This book stresses planned preventive maintenance actions as being of greater importance in our discussion. However, we do not exclude the necessity of eliminating failures during the execution of a planned repair. This situation may very well arise when an unsignalled failure is discovered in the course of a planned action such as inspection, for example.

The very possibility of varying the timing of the planned actions and their details (number of elements needing replacement, degree of renewal, etc.) provides for the selection of the best servicing policy.

2.2. Criteria of Optimality

2.2.1. Operation over a fixed period of time

Let us assume that the system is intended for operation during the finite time interval.

A useful criterion for this case would be the probability $R(t, z)$ that the system will be in an operational state at time t and will not fail during a subsequent period of time having the length z. This index is sometimes termed as a "strategic reliability" and is very important for specific "one-shot" military equipment. If the time z for completion of the military action is negligibly small, one may introduce another criterion $R(t, 0) = R(t)$ – the probability that the system will be operable at time t.

For technological systems it is more expedient to use criteria which reflect the economic indices of their work. During the total time of failure-free operation t, a random number of ER, PR and inspections will occur. Everyone of these actions is associated with certain losses expressed in time or in money. Therefore, for the time t a certain random amount V_t of time or cost will be accumulated. It is expedient to assume the ME of the value V_t, $\boldsymbol{M}\{V_t\} = \hat{V}_t$, as a quantitative characteristic of the system's performance. If all losses are dimensioned in units of time, then $\boldsymbol{M}\{V_t\} = \theta_t$ is a mean downtime of the system. For those systems whose effectiveness is proportional to the amount of the failure-free operation time, an important performance characteristic is the ratio

$$g_u(t) = t/(t + \theta_t) \tag{2.1}$$

called operational readiness (in some sources the term *coefficient of readiness* is used – see, for example, [68, p. 48]).

Another important criterion closely related to (2.1) is the rate of losses

$$g_t = \theta_t/t \tag{2.2}$$

(called in the cited source the *coefficient of preventive maintenance*).

2.2.2. Operation over an infinite period of time

If the system is intended to function during a period of time which is essentially longer than the mean failure-free operation time, then it is convenient to assume that the system is operating an infinitely long period of time. All criteria introduced for the case of a finite time interval should be naturally modified when $t \to \infty$. Denote

$$R(z) = \lim_{t\to\infty} R(t, z) \tag{2.3}$$

the probability of failure-free operation during time interval $(t, t+z)$ when $t \to \infty$. If in (2.3) we set $z = 0$, then the value $R_\infty(0)$ will express the probability that the system will be in an operating state at some moment t when t is large. $R_\infty(0)$ will be termed *operational readiness* (OR). Let us give another interpretation for OR. Figure 4 shows the alternation of random failure-free time periods $\mathbf{t}_i$ and random downtime periods $\mathbf{t}_i^*$ over a long time period. If $\mathbf{t}_i$ and $\mathbf{t}_i^*$ are independent random variables, which is usually a valid assumption, then it may be shown [19, Section 2.3] that

$$R_\infty(0) = \frac{\boldsymbol{M}\{\mathbf{t}_i\}}{\boldsymbol{M}\{\mathbf{t}_i\} + \boldsymbol{M}\{\mathbf{t}_i^*\}}, \tag{2.4}$$

i.e. $R_\infty(0)$ is equal to the ratio of the mathematical expectation (mean value)

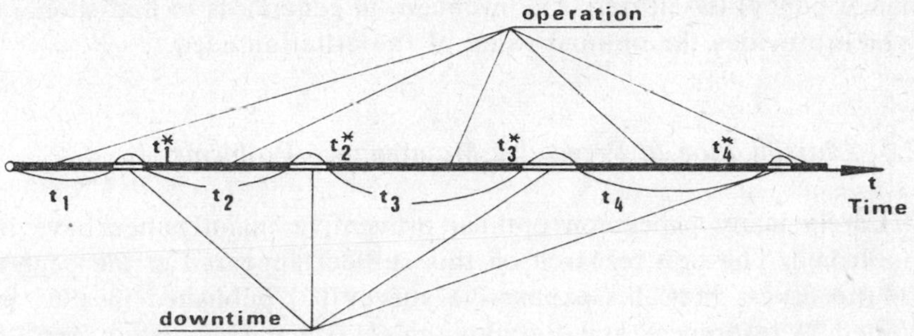

Figure 4. Operation- and down-time alternation in a long-term operating system.

(ME) of the failure-free operation time over the ME of one cycle consisting of the operation and post-operation downtime.

It can be shown that

$$\lim_{t\to\infty} g_u(t) = R_\infty(0).$$

Let us take a long period of time with n cycles of operation and n downtimes, and consider the ratio

$$g_n = \frac{\sum_{i=1}^{n} t_i}{\sum_{i=1}^{n} t_i + \sum_{i=1}^{n} t_i^*} = \frac{\sum_{i=1}^{n} t_i/n}{\sum_{i=1}^{n} t_i/n + \sum_{i=1}^{n} t_i^*/n}.$$

When we set $n \to \infty$, the expressions in the numerator and the denominator tend with probability 1 to $M\{t_i\}$ and $M\{t_i^*\}$ and g_n by $n\to\infty$ tends with probability 1 to $R_\infty(0)$. From this it follows that OR is equal to the fraction of time when the system is in the operational state (failure-free). Situation with criteria expressing the average losses $\hat{V}_t$ are more difficult because, formally speaking, these losses do not remain finite when $t\to\infty$.

In many cases, however, there is a way out because the value $\hat{V}_t$ is approximately proportional to t when t tends to infinity. In that case, one can speak about the rate of losses per unit time

$$g = \lim_{t\to\infty} \hat{V}_t/t. \tag{2.5}$$

All criteria introduced in this section take specific values when the properties of the system are known and the specific service rule (maintenance policy) is selected. The problem, in general, is to find such a rule which provides the optimal value of the criterion used.

2.3. Classification of Preventive Maintenance Problems

Lately many papers on optimal preventive maintenance have been published. The first research on this subject appeared at the beginning of the fifties. McCall's exhaustive survey [61] published in 1965 mentioned 88 references and Barzilovitch's [13] and [14], dated 1966–1967, more than two hundred. It would be desirable to have a classification of preventive maintenance problems which could serve as a guide to the "flow" of published works which has, apparently, been increasing greatly.

Classifications can be made on different bases. We suppose that the most expedient principle would be to classify all problems according to the statistical properties of the maintained systems or elements. Let us consider two situations.

(1) The *a priori* information or the data received during the system's operation permits us to distinguish only two states of the object under consideration: the failure and nonfailure states (see Examples A and B in the previous section). Most typical is knowledge of the DF of the failure-free operation time of the object. The preventive maintenance in this case is based on the data on DF and, of course, on the time which has elapsed since the last repair took place. It may happen that we know only some parameters or moments of the failure-free operation time, for example, the mean value and the variation, or an *a priori* distribution of those parameters. For this level of information two kinds of approaches may be proposed. The first one is so-called minimax: the best preventive maintenance policy is sought for the "worst" DF having the prescribed values of parameters or moments. The second approach is called an adaptive one. On the basis of the supposition about *a priori* distribution of the unknown parameters, an initial maintenance policy is chosen; acquiring information about the unknown parameters helps to revise the initial prior distribution and to select the optimal preventive maintenance policy.

(2) The *a priori* information and the data received during the operation enables us to define several states of the object which are "intermediate" between "the system is quite new" and "the system is in a failure state". Such situations take place when the capacity for work is determined by

some randomly changing parameter $\eta(t)$ available for observation (see Example C in the previous section). Sometimes the output characteristics of the system serve as such a parameter. In several cases we can define the parameter artificially. If there is a system with standby redundancy, it is possible to consider the number of failed standby elements at time t as a parameter $\eta(t)$ which gives information about the state of the system.

It is supposed that the parameter $\eta(t)$ can have more than two levels; otherwise we would be in the previous situation where one level corresponds to the nonfailure state and the other one to the failure state.

Thus, a second class of preventive maintenance situations is created where decisions on the state of the system are based on the observation of some randomly changing parameter which has more than two levels. We shall refer to this class of problems as a preventive maintenance based on the system parameter supervision.

Similar to case 1, one may have a situation when the statistical description of the parameter $\eta(t)$ is not complete. We have discussed (in Example D of the previous section) the model where the process $\eta(t)$ – the precision of a machine – was not measurable but another random process (the number of defective articles in the sample) stochastically connected with the first one was observed. All models where the maintenance policy is based on the indirect observation of the operating systems parameter are usually very complicated for mathematical treatment.

Therefore, we distinguish two main classes of preventive maintenance problems: one based on the information concerning only the DF of time-to-failure and the second based on the observation of some random parameter of the system. Within the framework of these two classes it is possible to solve the problem for a finite and infinite time interval. Usually, these two cases (termed as nonstationary and stationary) differ significantly in their mathematical treatment.

It must be said that the above description is neither comprehensive nor the only one possible. One may continue the classification by involving such properties of the system as methods used for failure discovery (there is a failure signalling system or not), or removal characteristics (does the removal renew the system entirely or not), etc. We shall not dwell on these questions because they are not important for our exposition.

3. Definition and Basic Properties of the Controlled Semi-Markov Process

3.1. Definition of the SMP

We have already mentioned that it is possible to interpret the problems of choosing optimal preventive maintenance actions in terms of finding the optimal control of a random process. It seems to us that a very convenient tool for our purpose is the semi-Markov process (SMP).† In terms of SMP it is possible to give satisfactory descriptions of a broad circle of optimal preventive maintenance models. At the same time, there exist very effective algorithms for these processes which might be used for finding the optimal service policy.

To make the concept of a SMP clearer we shall start with an elementary example which demonstrates the relationship between the finite-state Markov chain and the SMP.

Let us consider a chain with three states. The matrix of transition probabilities is

$$\tilde{\mathbf{P}} = \begin{Vmatrix} \tilde{p}_{11} & \tilde{p}_{12} & \tilde{p}_{13} \\ \tilde{p}_{21} & \tilde{p}_{22} & \tilde{p}_{23} \\ \tilde{p}_{31} & \tilde{p}_{32} & \tilde{p}_{33} \end{Vmatrix}.$$

This chain may be considered as a process developing in time, in the following manner.

Imagine a fictitious particle which occupies at any moment one of three levels denoted E_1, E_2 and E_3 (in other words, the particle can be in three states). At instant $t = 0$, the particle is placed, for example, in E_1. At instant $t = 1$, it jumps with the probability $\tilde{p}_{1j}$ into state E_{j_1}; $1 - \tilde{p}_{12} - \tilde{p}_{13}$ is the probability that the particle "jumps" into the initial state E_1, that is $E_{j_1} = E_1$. Analogously, the particle moves at time $t = 2$ from the state E_{j_1} into a state E_{j_2}, and so on.

A trajectory of this particle is plotted in Figure 5. A case is shown where the particle was three time units on level E_1 (that is, twice it jumped from E_1 into E_1), two time units on the level E_2, and so on. If the particle is in state E_i, then sooner or later it will leave this state and go to state E_j, $j \neq i$ (of course, it is supposed that $\tilde{p}_{ii} < 1$). Let us denote the event "the particle spent k time units in state E_i and later on entered state E_j, $j \neq i$"

†The bibliography of SMP now lists hundreds of works. We mention only a few: Howard[47] and [48, Vol. 2], Jewell[50], Pyke[69] and [70].

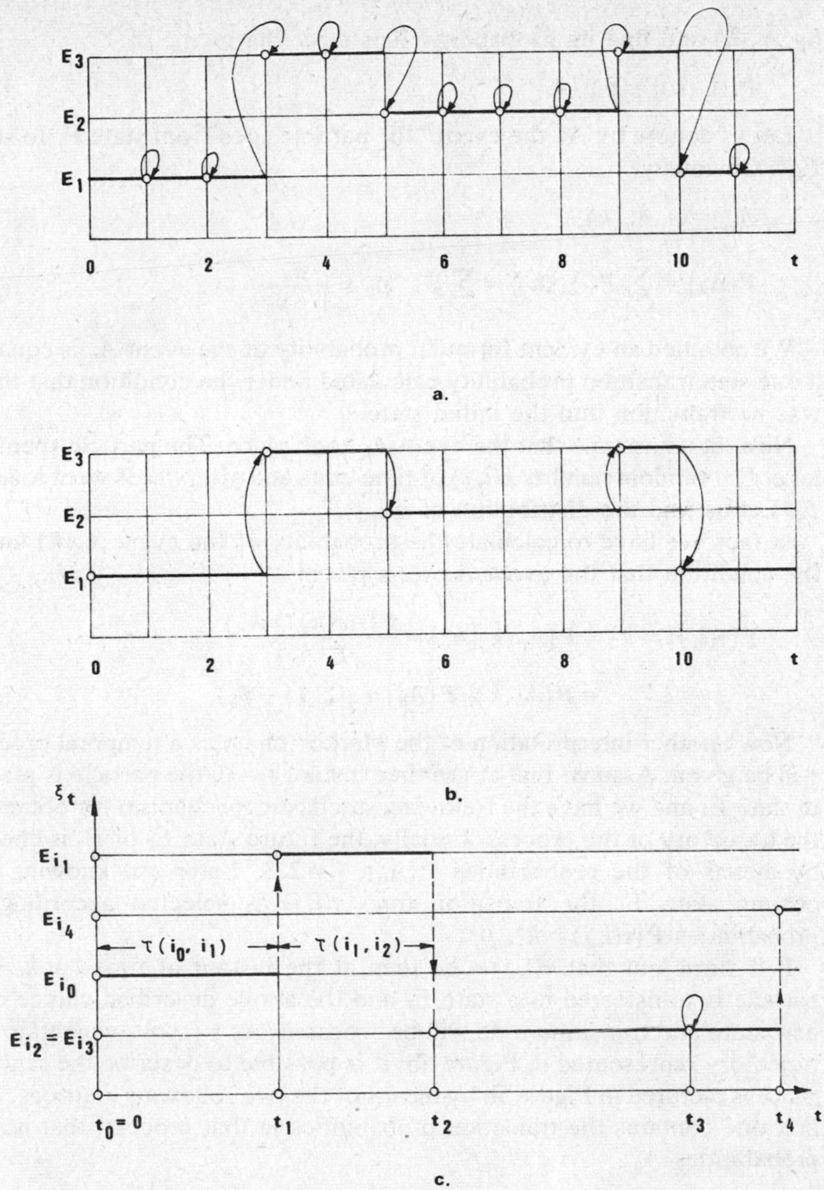

Figure 5. (a) Trajectory of a Markovian particle. (b) Sample function of a random process obtained from (a) concerning only transitions of type $E_i \to E_j$, $j \neq i$. (c) Sample function of a semi-Markov process.

by $A_{ij}(k)$ and find its probability. It is clear that

$$P\{A_{ij}(k)\} = \tilde{p}_{ii}^{\,k-1}\tilde{p}_{ij}. \tag{3.1}$$

Let us denote by A_{ij} the event "the particle goes from state E_i to state E_j". Obviously,

$$A_{ij} = \bigcup_{k=1}^{\infty} A_{ij}(k),$$

$$P\{A_{ij}\} = \sum_{k=1}^{\infty} P\{A_{ij}(k)\} = \sum_{k=1}^{\infty} \tilde{p}_{ii}^{\,k-1}\tilde{p}_{ij} = \frac{\tilde{p}_{ij}}{1-\tilde{p}_{ii}}. \tag{3.2}$$

We obtained an evident formula: probability of the event A_{ij} is equal to a one-step transition probability calculated under the condition that there was no transition into the initial state.

Now let us assume that the event A_{ij} took place. The particle spent on level E_i a random number $\tau(i, j)$ of time units and afterwards went to level E_j. Let us find the distribution of $\tau(i, j)$.

In fact, we have to calculate the probability of the event $A_{ij}(k)$ under the condition that the event A_{ij} took place:

$$P\{\tau(i, j) = k\} = P\{A_{ij}(k) | A_{ij}\} = \frac{P\{A_{ij}(k) \cap A_{ij}\}}{P\{A_{ij}\}}$$

$$= P\{A_{ij}(k)\}/P\{A_{ij}\} = \tilde{p}_{ii}^{\,k-1}(1-\tilde{p}_{ii}). \tag{3.3}$$

Now another interpretation of the Markov chain as a temporal process will be given. Assume that at the time instant $t = 0$, the particle is placed in state E_1 and we have the following stochastic mechanism for obtaining the trajectory of the process. Initially, the future state E_2 or E_3 is chosen by means of the probabilities $P\{A_{ij}\}$, $j = 2, 3$. Later on, knowing the coming state E_j, the transition time $\tau(i, j)$ is selected according to probabilities $P\{\tau(i, j) = k\}$.

If it turns out that $\tau(i, j) = k_0$, then at the instant of time $t = k_0$, the particle is transferred into state E_j and the above described choice of a new state and transition time will be repeated. As a result, we obtain the trajectory represented in Figure 5b. It is possible to describe the random process pictured in Figure 5b by means of the two following matrices. The first one contains the transition probabilities in that process, that is, the probabilities A_{ij}:

$$\mathbf{P} = \begin{Vmatrix} 0 & p_{12} & p_{13} \\ p_{21} & 0 & p_{23} \\ p_{31} & p_{32} & 0 \end{Vmatrix},$$

$$p_{ij} = \boldsymbol{P}\{A_{ij}\} = \frac{\tilde{p}_{ij}}{1-\tilde{p}_{ii}}, \qquad j \neq i. \tag{3.4}$$

The second matrix contains the probability DFs of the random transition times $\tau(i, j)$ calculated under the condition that these transitions took place:

$$\mathbf{F} = \|\boldsymbol{P}\{\tau(i, j) \leq k\}\|, \qquad i, j = 1, 2, 3; \quad i \neq j; \quad k = 1, 2, \ldots. \tag{3.5}$$

The distribution $\boldsymbol{P}\{\tau(i, j) = k\}$ follows immediately from (3.3). Instead of these two matrices, we can introduce only one matrix **Q**, which will contain all the necessary information for obtaining matrices **P** and **F**. Define **Q** as follows:

$$\mathbf{Q} = \|\boldsymbol{P}\{A_{ij}(k) \leq l\}\|, \qquad j \neq i; \quad i, j = 1, 2, 3; \quad l = 1, 2, \ldots,$$

where

$$\{A_{ij}(k) \leq l\} = \bigcup_{k=1}^{l} A_{ij}(k).$$

Indeed, from (3.1) it follows that

$$\boldsymbol{P}\{A_{ij}(k) \leq l\} = \tilde{p}_{ij}(1 - \tilde{p}_{ii}^{\,l})/(1 - \tilde{p}_{ii}).$$

Then

$$\boldsymbol{P}\{A_{ij}(k) \leq l\} \xrightarrow[l \to \infty]{} \boldsymbol{P}\{A_{ij}\} = \tilde{p}_{ij}/(1 - \tilde{p}_{ii}) = \tilde{p}_{ij}.$$

Proceeding from **Q**, it is easy to find the distribution of $\tau(i, j)$:

$$\boldsymbol{P}\{\tau(i, j) \leq l\} = \boldsymbol{P}\{A_{ij}(k) \leq l\}/\boldsymbol{P}\{A_{ij}\}.$$

The random process defined by means of matrices (3.4) and (3.5) has three peculiarities inherited from the initial Markov chain $\tilde{\mathbf{P}}$. The first one, called a Markovian property, is that the evolution of the process after the instant t (if it is known that at this instant a transition into state E_i took place) does not depend on the past of the process, i.e., on which transitions took place before the instant t and when.

The second peculiarity stipulates the specific form of the distribution of the random variables $\tau(i, j)$: according to (3.3), $\tau(i, j)$ has a geometrical distribution, which does not depend on the index j.

Finally, we notice that the values on the diagonal of the matrix (3.4) are zeros. This means that the transitions of type $E_i \to E_i$ are excluded. It is a natural consequence of the way the process was constructed: we defined the transition only when in the initial chain the transition $E_i \to E_j$, $j \neq i$, took place.

Now let us try to generalize our process by retaining the Markov property and ignoring the specific form of the distribution of the transition time $\tau(i, j)$. It will be assumed, as before, that after the transition from E_k to E_i, a new state is selected according to a collection of transition probabilities $\{p_{ij}\}$, $p_{ij} \geq 0$, $\Sigma_j\, p_{ij} = 1$, independently of the behaviour of the process before the transition into E_i was made. Later on, the particle remains in state E_i a random time $\tau(i, j)$ having an *arbitrary* distribution function $F_{ij}(t)$.

In our new process, we shall not demand that the transition always lead to a change of the occupied state. It will be assumed that with probability p_{ii}, the particle is moving to its initial state E_i and that such a "transition" lasts a random time $\tau(i, i)$ having DF $F_{ii}(t)$. In this way, we have arrived at the semi-Markov process. We shall now give a more formal description of it and afterwards discuss several examples.

Consider a random process ξ_t with a finite set of states $\mathbf{E} = \{E_1, \ldots, E_n\}$.† The transitions in that process occur at random time instants $\mathbf{t}_1, \mathbf{t}_2, \ldots$. It will be assumed that initially the transition into E_{i_0} took place at time $t_0 = 0$: $\xi_0 = E_{i_0}$. Denote by $\boldsymbol{I}_m$ the state which will be entered at the moment $\mathbf{t}_m$, $m = 1, 2, \ldots$. It is assumed that

$$\boldsymbol{P}\{\boldsymbol{I}_m = E_j \,|\, \boldsymbol{I}_0, \boldsymbol{I}_1, \ldots, \boldsymbol{I}_{m-1} = E_i\} = \boldsymbol{P}\{\boldsymbol{I}_m = E_j \,|\, \boldsymbol{I}_{m-1} = E_i\} = p_{ij}, \tag{3.6}$$

or, in other terms, the states $\boldsymbol{I}_m$ form a time-homogeneous Markov chain with the matrix of transition probabilities

$$\mathbf{P} = \|p_{ij}\|. \tag{3.7}$$

Let the transition into state E_i occur at time instant $\mathbf{t}_r$ and the next transition in state E_j at time instant $\mathbf{t}_{r+1}$. Denote $\mathbf{t}_{r+1} - \mathbf{t}_r = \tau(i, j)$. The functions $F_{ij}(t)$ will be termed transition distributions. In other words,

$$F_{ij}(t) = \boldsymbol{P}\{\tau(i, j) \leq t\}. \tag{3.8}$$

To every pair of indices, i, j ($p_{ij} > 0$), there is one corresponding DF $F_{ij}(t)$. (It is assumed that $F_{ij}(t) = 0$ for $t < 0$.) $F_{ij}(t)$ is the DF of the *transition from* E_i *headed towards* E_j.

Obtaining one trajectory of the process ξ_t may be described in the following way (see Figure 5c). Initially, a certain state E_{i_0} is taken: $\boldsymbol{I}_0 = E_{i_0}$. Then according to the matrix (3.7) the next state E_{i_1} is selected. After the

†See also Appendix 1, Definition of the SMP.

new state is known, one value of the random variable $\tau(i_0, i_1)$ from the population with DF $F_{i_0 i_1}(t)$ is selected. The process ξ_t occupies state E_{i_0} time $\tau(i_0, i_1)$ and later on jumps into state E_{i_1}. Then the described procedure of selection of the future state E_{i_1} and the holding time $\tau(i_1, i_2)$ continues, and so on. It is essentially important that time $\tau(i_m, i_{m+1})$ does not depend on the "biography" of the process before entering state E_{i_m}.

The time between adjacent transitions will be called a "step", and describing the process we shall use the words "the process stayed in state E_{i_0} during time $\tau(i_0, i_1)$, went to state E_{i_1}", and so on. We use the notation

$$\mathbf{F} = \|F_{ij}(t)\| \tag{3.9}$$

for the matrix of transition DFs and assume that if for a certain pair of indices i, j, $p_{ij} = 0$, then $F_{ij} \equiv 0$.

One may define a semi-Markov process by means of one matrix $\mathbf{Q}$

$$\mathbf{Q} = \|Q_{ij}(t)\|, \qquad Q_{ij}(t) = p_{ij}F_{ij}(t), \tag{3.10}$$

which determines simultaneously the bivariate distribution of the future state and the transition interval. Indeed, $Q_{ij}(t)$ is a probability of the event "given the initial state E_i process ξ_t will go to state E_j and the holding time (i.e. the transition time) will not exceed the value t". Given $Q_{ij}(t)$, it is possible to find p_{ij} and $F_{ij}(t)$ because $Q_{ij}(\infty) = p_{ij}$ and $F_{ij}(t) = Q_{ij}(t)/Q_{ij}(\infty)$.

Usually two types of random variables of the transition time $\tau(i, j)$ will appear in the sequel: the continuous and the discrete one. In the first case, it will be assumed that DF $F_{ij}(t)$ has a density function, that is, there is a function $f_{ij}(t)$ such that $F_{ij}(t) = \int_0^t f_{ij}(x)\,dx$. In the second case, the variable $\tau(i, j)$ can take with nonzero probabilities only a finite or numerable set of values. Of course, constant value is a particular case of a discrete-type random variable.

In the sequel, expressions for the ME of values $\tau(i, j)$ or of some functions from these values $R(\tau(i, j))$ will often be used. Since the type of random variable is not known beforehand, the notation in the form of the Stieltjes integral will be used:

$$M\{R(\tau)\} = \int_0^\infty R(x)\,dF(x).$$

If the random variable is of a "mixed" type and its DF has jumps at points $\theta_1, \theta_2, \theta_3, \ldots$, and apart from these points there exists a density

$f(t)$, then the above notation has the following meaning:

$$M\{R(\tau)\} = \sum_i \int_{\theta_i}^{\theta_{i+1}} R(z)f(z)\,dz + \sum_k p_k R(\theta_k).$$

Let us now discuss a few examples of SMP related to preventive maintenance.

Example 1. *The "work-repair" operation scheme.* Assume that process ξ_t has the matrices **P** and **F** as follows:

$$\mathbf{P} = \begin{Vmatrix} 0 & 1 \\ 1 & 0 \end{Vmatrix}, \qquad \mathbf{F} = \begin{Vmatrix} 0 & G_1(t) \\ G_2(t) & 0 \end{Vmatrix}. \tag{3.11}$$

The states alternate $E_1 \to E_2 \to E_1 \to E_2 \to \cdots$ (see Figure 6a). This process permits a simple interpretation. At time $t_0 = 0$, the element is engaged at work, state E_1 is a state of failure-free operation; at the random moment

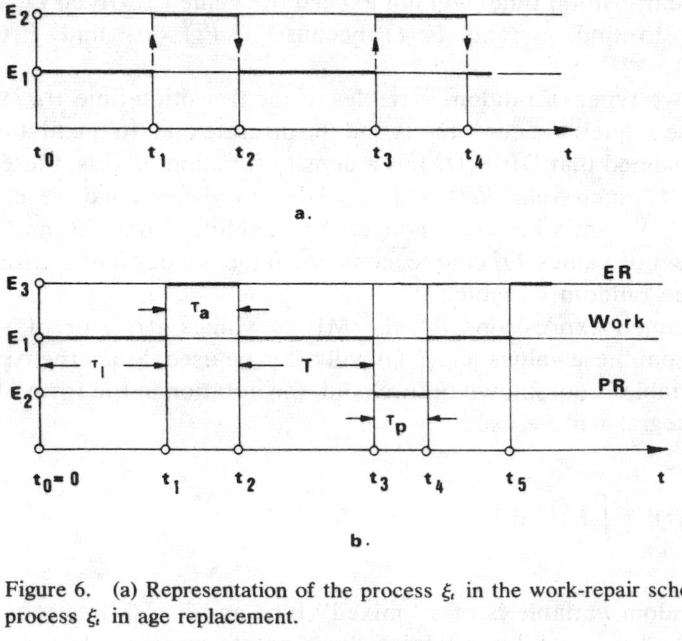

Figure 6. (a) Representation of the process ξ_t in the work-repair scheme. (b) The process ξ_t in age replacement.

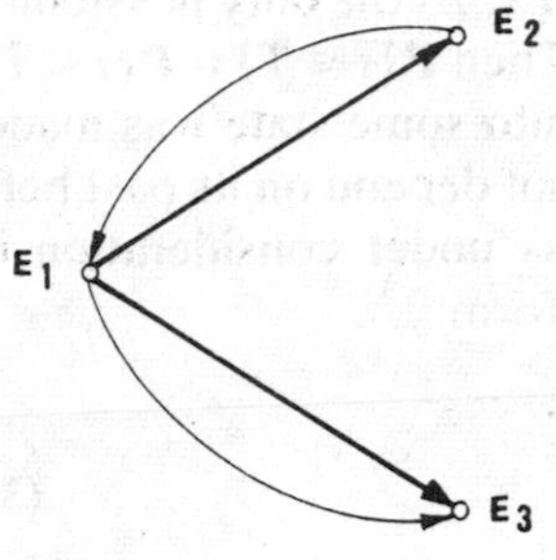

Figure 6 (*continued*). (c) Transition diagram in age replacement.

t_1, a failure appears and the element passes to state E_2 (which is the state of repair); at time t_2, the repair is finished and the element starts to work, and so on. The intervals of failure-free operation times and repair lengths are independent random variables; according to (3.11), their DFs are respectively $G_1(t)$ and $G_2(t)$.

Example 2. *Age replacement.* We consider the following kind of element's repair. At time $t = 0$, a new element is starting to work. If during the time span T of continuous operation, the element does not fail, then at time instant $t = T$, the element will be replaced by a new one (we refer to this action as PR). Otherwise, at the moment when a signal denotes failure, the ER starts. Both ER and PR last a random time and their DFs are $F_p(t)$ and $F_a(t)$ respectively. The element itself has a DF of time-to-failure $F(t)$. Every repair completely renews the element; it may be assumed that the element is always replaced by a brand new one. If the renewal of the element was finished at time instant t_1, the element starts to work, and the next PR is planned at the time moment $t_1 + T$, and so on. Therefore, "compulsory" measures (PR) are executed if and only if the age of the element reaches the "critical" value T.

We can say that the element can be in three different states: failure-free operation, preventive repair or emergency repair. Denote these states E_1, E_2 and E_3 respectively. Assume that the element failed at the instant $t_1 < T$, was repaired during time τ_a, worked without failure during time span T, went to PR, and so on. The sample function of the random process corresponding to this description is given in Figure 6c.

Let us clear up the question about the stochastic mechanism governing the transitions $E_i \to E_j$ in this process. If the failure-free time τ is less than

T, the transition $E_1 \to E_3$ occurs. From states E_2, E_3 the only possibility is to go to state E_1. Assume that $\boldsymbol{P}\{\boldsymbol{\tau} = T\} = 0$. Then $\boldsymbol{P}\{\boldsymbol{\tau} \leqslant T\} = \boldsymbol{P}\{\boldsymbol{\tau} < T\} = F(T)$. It should be noted that if a transition into some state was made at instant t, then the future of the process does not depend on its past before time t. From this it follows that the process under consideration is a semi-Markovian one. The matrix $\mathbf{P}$ has the form:

$$\mathbf{P} = \begin{Vmatrix} 0 & 1-F(T) & F(T \\ 1 & 0 & 0 \\ 1 & 0 & 0 \end{Vmatrix}. \tag{3.12}$$

It is obvious that the transition $E_1 \to E_2$ always lasts a constant time T. Therefore,

$$F_{12}(t) = \begin{cases} 0, & t < T; \\ 1, & t \geqslant T. \end{cases}$$

DF $F_{13}(t)$ defines the probability of a transition $E_1 \to E_3$ which lasts less than time t; this is a conditional probability calculated on the assumption that this transition indeed took place. Therefore,

$$F_{13}(t) = \begin{cases} F(t)/F(T), & t < T; \\ 1, & t \geqslant T. \end{cases}$$

Later on, it is clear that

$$F_{31}(t) = F_a(t), \qquad F_{21}(t) = F_p(t).$$

The example considered is a typical one. In the sequel, we shall repeatedly use the description of the maintained systems' behaviour in terms of a SMP

Example 3. *Continuous-time Markov chain as a special case of SMP.* We return to the Markov chain described at the beginning of this section. Let us assume that the matrix $\mathbf{P}$ describes the transitions occurring every short time interval Δt. It is natural to require that the transition probabilities from E_i to E_j, $j \neq i$, during this short time period should also be small values. This demand is represented in the following form:

$$\tilde{p}_{ij} \underset{\Delta t \to 0}{=} \lambda_{ij}\Delta t + o(\Delta t), \qquad j \neq i. \tag{3.13}$$

Omitting the term $o(\Delta t)$, we can write the matrix $\tilde{\mathbf{P}}_{\Delta t}$ of transition probabilities during the short time interval Δt in the following form:

$$\tilde{\mathbf{P}}_{\Delta t} = \begin{Vmatrix} 1-(\lambda_{12}+\lambda_{13})\Delta t & \lambda_{12}\Delta t & \lambda_{13}\Delta t \\ \lambda_{21}\Delta t & 1-(\lambda_{21}+\lambda_{23})\Delta t & \lambda_{23}\Delta t \\ \lambda_{31}\Delta t & \lambda_{32}\Delta t & 1-(\lambda_{31}+\lambda_{32})\Delta t \end{Vmatrix}.$$

Assume that the particle went at time instant t to state E_1. Let us find the probability of the event $A_{ij} = \{\text{the next transition will be into state } E_j\}$. According to (3.2),

$$\boldsymbol{P}\{A_{ij}\} = \lambda_{1j}/(\lambda_{12}+\lambda_{13}) + o(\Delta t), \qquad j = 2, 3. \tag{3.14}$$

Assume that it is known that the transition $E_1 \to E_j$, $j \neq 1$, took place and that it occurred at time $\mathbf{t}_1 = t + \boldsymbol{\tau}_{1j}$. Let us find the distribution of the random variable $\boldsymbol{\tau}_{1j}$. In this case, it is more convenient to operate with the distribution in the integral form. By (3.3)

$$\begin{aligned} \boldsymbol{P}\left\{\boldsymbol{\tau}(1, j) \leqslant k = \frac{t}{\Delta t}\right\} &= \sum_{n=1}^{k} \tilde{p}_{11}^{n-1}(1-\tilde{p}_{11}) = 1 - \tilde{p}_{11}^{k} \\ &= 1 - [1-(\lambda_{12}+\lambda_{13})\Delta t]^{t/\Delta t} + o(\Delta t) \xrightarrow[\Delta t \to 0]{} \\ &\quad 1 - \exp[-t(\lambda_{12}+\lambda_{13})]. \end{aligned}$$

Similar expressions may be obtained for other states. If a SMP with states E_1, E_2 and E_3 is considered, then the matrices **P** and **F** will have the form

$$\mathbf{P} = \|p_{ij}\|, \qquad p_{ij} = \lambda_{ij}\Big/\sum_{j\neq i}\lambda_{ij}, \quad i \neq j; \qquad p_{ii} = 0;$$

$$\mathbf{F} = \|F_{ij}(t)\|, \qquad F_{ij}(t) = 1 - \exp\left(-\sum_{j\neq i}\lambda_{ij}t\right), \qquad i \neq j.$$

The random process with transition probabilities given by (3.13) is called a continuous-time Markov chain. The coefficients λ_{ij} will be referred to here as transition rates. From our reasoning, it follows that the evolution of this process is the following: the particle stays in state E_i a random exponentially distributed time and later on jumps into state E_j, $j \neq i$, with the probability $\lambda_{ij}/\Sigma_{j\neq i}\lambda_{ij}$.

Up to this point in all the examples considered, the matrix **P** had zeros on its diagonal, that is, the transitions $E_i \to E_i$ were excluded. Now we consider an example in which this peculiarity does not take place.

Example 4. *Age replacement, momentary repair.* Let us return to Example 2 (Age replacement) and assume that all repairs take a negligible amount of time. In a formal sense this is equivalent to the assumption that only failure-free time spans are placed on the time axis and that downtime periods are moved off. We define two states: E_1 – the element is operating, operation starting after ER; E_2 – the element is operating, operation starting after PR. The sequence

$$E_1 \to E_1 \to E_2 \to E_1 \to$$

means that at time $t = 0$, ER takes place; later on the element failed before time $t = T$, i.e. ER was again made; the next operation period of length T was failure-free, and so on. In this situation the matrix $\mathbf{P}$ has the form

$$\mathbf{P} = \begin{Vmatrix} F(T) & 1 - F(T) \\ F(T) & 1 - F(T) \end{Vmatrix},$$

and

$$F_{21}(t) = F_{11}(t) = \begin{cases} F(t)/F(T), & t < T, \\ 1, & t \geqslant T, \end{cases}$$

$$F_{12}(t) = F_{22}(t) = \begin{cases} 0, & t < T, \\ 1, & t \geqslant T. \end{cases}$$

3.2. *Control Actions and Types of Strategies*

Let us proceed again with the consideration of Example 2 of Section 3.1. In this example, a stochastic mechanism governing the random process ξ_t was given by means of matrices (3.12) and the collection of DFs $F_{ij}(t)$. According to the description of the example, this mechanism remained unchanged over the entire operation period. Now let us imagine the following situation. There is a possibility of selecting different values of the "critical" age T, assuming them to be equal to T_1, or T_2. Next, let there be a choice in the repair "technology" which manifests itself as a choice between two possible DFs, $F_p^*(t)$ or $F_p^{**}(t)$ for the length of the preventive repair, and between $F_a^*(t)$ or $F_a^{**}(t)$ for the emergency repair.

It is convenient to introduce the following terminology. Let us say that in states E_i it is possible to carry out one of several available *control actions* (CAs). Thus, in state E_1, the control actions T_1 or T_2 are available; in state E_2 one can select $F_p^*(t)$ or $F_p^{**}(t)$; for state E_3 there is a choice between $F_a^*(t)$ and $F_a^{**}(t)$.

The choice of a CA is executed immediately after the transition in the process ξ_t has occurred.

Let us suppose that the mth transition was made at random time $\mathbf{t}_m$ and the system passed to state E_{i_m}. The choice of the CA made at time $\mathbf{t}_m$ determines the stochastic mechanism governing the process ξ_t during one step "in advance". In other words, the choice of a CA determines the probability of transition into the next state $E_{i_{m+1}}$ at a certain random moment $\mathbf{t}_{m+1}$. Thus, the choice of time T_2 as a critical age for state E_1 determines the following probabilities for the transition into states E_2 and E_3:

$$p_{12} = 1 - F(T_2), \qquad p_{13} = F(T_2).$$

The rule which gives a prescription for taking control actions for all states over the entire functioning period is referred to as a "strategy", for which the notation $\boldsymbol{\sigma}$ is used.

We proceed with consideration of several typical situations.

A. For each state E_i there is one CA always associated with it. The tuple of CAs

$$\boldsymbol{\sigma} = (\sigma_1, \sigma_2, \ldots, \sigma_n) \tag{3.15}$$

denotes completely the stochastic behaviour of the system under consideration.

In the above example, we can decide that the following CAs would be selected: $F_a(t) = F_a^*(t)$, $F_p(t) = F_p^{**}(t)$, $T = T_2$. The strategy is, therefore, the following:

$$\boldsymbol{\sigma} = (T_2, F_p^{**}, F_a^*).$$

In this example there exist $2^3 = 8$ different strategies of that kind, and each one determines completely the probabilistic behaviour of the system. We shall refer to the strategy of the form (3.15) as a Markovian stationary strategy of the first type. (The term "Markovian" will be discussed below.)

B. The choice of a certain CA σ_i after the transition into state E_i at the moment $\mathbf{t}_m$ depends on the number m of that transition: $\sigma_i = \sigma_i(m)$, $i = 1, \ldots, n$. The strategy has the form

$$\boldsymbol{\sigma} = \boldsymbol{\sigma}(m) = \{\sigma_1(m), \sigma_2(m), \ldots, \sigma_n(m)\}, \qquad m = 0, 1, 2, \ldots. \tag{3.16}$$

Thus, in this example, we can prescribe the following rule: take for the first 10 steps (i.e. at time instants $t_0, t_1, t_2, t_3, \ldots, t_9$) strategy $\boldsymbol{\sigma} = (T_1, F_p^*, F_a^*)$ and afterwards strategy $\boldsymbol{\sigma}' = (T_2, F_p^{**}, F_a^{**})$. It is obvious that the stochastic mechanism depends on the number of the step: the first ten transitions are governed by matrices **P** and **F** containing $T = T_1$, $F_p = F_p^*$ and $F_a = F_a^*$ and during the subsequent steps by the matrices with $T = T_2$, $F_p = F_p^{**}$ and $F_a = F_a^{**}$.

The strategy (3.16) will be referred to as a Markovian nonstationary with respect to the number of the step, or as the strategy of the second type. It is characteristic that by applying this strategy, the sequence of states E_{i_k}, $k = 0, 1, 2, \ldots$, forms a Markov chain which is nonhomogeneous with respect to the number of the step.

C. The choice of a certain CA σ_i for state E_i depends on the actual time when the decision is made, that is, on time t of the transition: $\sigma_i = \sigma_i(t)$. The strategy is a vector function depending on the time:

$$\underset{t\geq 0}{\boldsymbol{\sigma}} = \boldsymbol{\sigma}(t) = \{\sigma_1(t), \sigma_2(t), \ldots, \sigma_n(t)\}. \tag{3.17}$$

In our example, the following performance prescription may be proposed:

$$\boldsymbol{\sigma}(t) = \begin{cases} (T_1, F_p^*, F_a^*), & t \leq \hat{t}; \\ (T_2, F_p^{**}, F_a^{**}), & t > \hat{t}. \end{cases} \tag{3.18}$$

In other words, all CAs made up to time instant $\hat{t}$ correspond to the first row of (3.18), and all CAs made after this instant, to the second row.

A strategy similar to (3.17) will be referred to as a Markovian nonhomogeneous with respect to the time, or a strategy of the third type.

Let us summarize our description. The control means the following: let the system at the moment $\mathbf{t}_m$ enter state E_{i_m}; exactly at this time instant, a certain CA σ_{i_m} is selected which determines the statistical mechanism of the evolution of the process during one step. Based on σ_{i_m}, a next state E_j is selected as a sample from the conditional probabilities $p_{i_m,j}(\sigma_{i_m})$; the sojourn time until that next state is entered, $\boldsymbol{\tau}(i_m, j; \sigma_{i_m})$, is selected as a sample from the distribution $F_{i_m,j}(t; \sigma_{i_m})$.

It is essential that the rule prescribing the choice of the CA for a given state depends on the number of the transition (i.e. the number of the step) and on the value of time when the CA is selected.

In a general case, the rule which determines the choice of the CA may take into consideration not only the actual state, time and (or) number of the transition but also the entire "history" of the process up to the moment of making the decision.

Let us imagine, for example, the following "exotic" repair rule. The critical age (we refer to the example dealing with age replacement) is taken as $T = T_1$, if before the moment of the decision there were more ERs than PRs and, in the opposite case, the critical age $T = T_2$.

We are restricting ourselves in this book to a consideration only of control rules termed as Markovian which have the property described below.

The choice of the CA σ_{i_m} for the state E_{i_m} at the moment t_m *does not depend on the behaviour of the process* ξ_t *up to the moment* t_m; it may depend on the state E_{i_m} itself, on the number of the step m or on the time t_m. Formally, this property which declares the independence of the past can be written as follows. If the symbol $\xi^{(t_m)}$ represents the past of the process ξ_t up to the time t_m, then

$$\begin{aligned}\sigma_{i_m} &= f(\xi^{(t_m)}, E_{i_m}, m, t_m) \\ &= f(E_{i_m}, m, t_m).\end{aligned}$$

This independence of the previous "history" by taking the CA is what is called the Markovian property of the strategy.

The set of all CAs σ_i available in the state E_i is denoted by S_i: $\sigma_i \in S_i$. For the strategies of the first type, sets S_i are forever fixed. For considering the second and third type strategies we shall assume that these sets can depend on the number of the step or on the time:

$$\sigma_i(m) \in S_i(m), \qquad \sigma_i(t) \in S_i(t).$$

It will be assumed also that all S_i are finite. The set of all strategies will be denoted by S.

3.3. Costs (Returns)

We have mentioned before that performing different service actions which are directed at avoiding failure or preventing it is associated with certain expenditures or demands a certain amount of time. This fact permits the following formal interpretation in terms of the controlled random process ξ_t.

It will be assumed that when the process is in state E_i headed towards E_j, a cost (or return) is accumulated according to an arbitrary function $R_{ij}(\tau(i,j); \sigma_i)$, depending on i, j, the transition time $\tau(i,j)$ and the CA σ_i selected when the system entered state E_i.

Thus, in Example 2, Section 3.1, we can introduce the following costs: c_a if a failure occurs; c_p if preventive maintenance takes place. The following function R_{ij} corresponds to these costs:

$$R_{ij}(\tau(i,j);\sigma_i)=\begin{cases}c_a, & i=1, j=3;\\ c_p, & i=1, j=2;\\ 0, & \text{otherwise.}\end{cases}$$

The total cost $2c_a + c_p$ will be accumulated for two ERs and one PR in the situation plotted in Figure 6b, because two transitions $E_1 \to E_3$ and one transition $E_1 \to E_2$ took place.

It should be noted that we speak about *costs* in those cases when the strategy sought is directed to *minimize* some kind of expenses. In several problems, we are looking for a strategy which *maximizes* certain performance characteristics of the system, for example, the average failure-free operation time. In this case it is more convenient to use the word "return" instead of "cost". The term chosen depends on the specific situation.

Several other terms are used in literature, such as penalties or losses (usually subjected to minimization) and rewards, incomes, payoffs (which have to be maximized).

The expected average cost (or return) accumulated in several steps or during a given time will be studied. The following example illustrates the meaning of the term "average accumulated cost".

Let us consider again the age replacement (see Example 2, Section 3.1). At $t=0$ a new element is engaged in the operation. The functioning is planned for three steps, that is, for the cycle "operation-repair-operation". For the sake of simplicity, we assume that the element has the following failure-free operation DF: with probability p it fails after $t_1=5$ hours after the beginning of work and with probability $q=1-p$ after $t_2=10$ hours. Two CAs are available at the moment the element starts to work: to perform the PR after $T_1=8$ hours or $T_2=12$ hours. The DF of the duration of ER and PR are fixed as $F_a(t)$ and $F_p(t)$ respectively and, therefore, there is no choice of CA in states E_2 and E_3. The one-step returns will be defined as follows:

$$R_{13}(\tau(1,3)) = k\tau(1,3) - c_a = 5\tau(1,3) - 10; \qquad R_{31} = 0;$$
$$R_{12}(\tau(1,2)) = k\tau(1,2) - c_p = 5\tau(1,2) - 5; \qquad R_{21} = 0.$$

The physical meaning of these returns is the following: for every unit of failure-free operation time a return $k=5$ is counted; a negative return of 10 units is associated with the ER and 5 with the PR. In that case we have a total of four different strategies:

$$\boldsymbol{\sigma}^{(1)} = \{\sigma_1(0) = T_1; \sigma_1(2) = T_1\};$$
$$\boldsymbol{\sigma}^{(2)} = \{\sigma_1(0) = T_2; \sigma_1(2) = T_1\};$$
$$\boldsymbol{\sigma}^{(3)} = \{\sigma_1(0) = T_2; \sigma_1(2) = T_2\};$$
$$\boldsymbol{\sigma}^{(4)} = \{\sigma_1(0) = T_1; \sigma_1(2) = T_2\}.$$

Thus, the notation $\boldsymbol{\sigma}^{(4)}$ denotes a strategy in which the first period of PR was appointed as T_1 and the second as T_2.

Let us take strategy $\boldsymbol{\sigma}^{(1)}$. This strategy defines the random mechanism which governs the process during three steps. We consider all possible trajectories in the state space E_1, E_2, E_3. It is easy to check that all possibilities are exhausted by four different trajectories (see Figure 7 which also presents the necessary explanations).

Since there are several strategies, it is desirable to know how to compare them. The more natural basis for the comparison would be the total expected cost (or return) accumulated during the prescribed number of steps. To calculate the expected cost it is necessary to know the probabilistic "weight" of the trajectory, and afterwards to calculate the

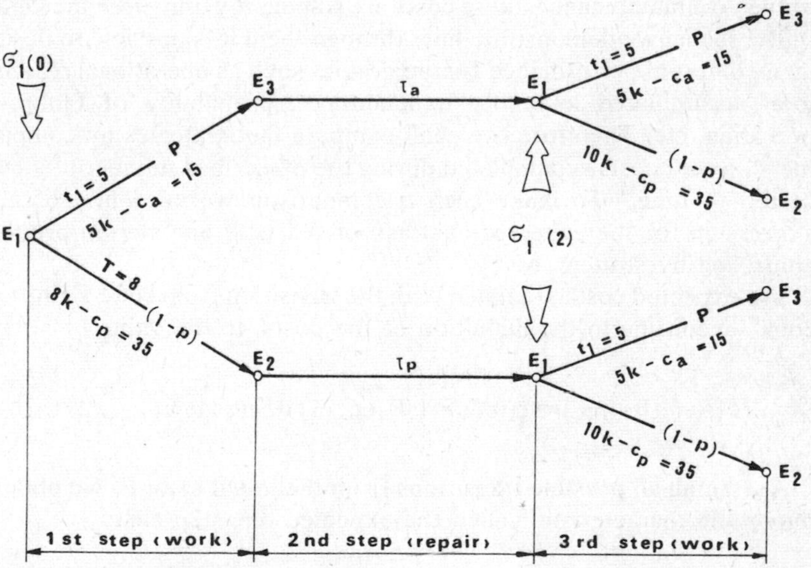

Figure 7. The "bundle" of trajectories in a three-step process describing age replacement under the strategy $\boldsymbol{\sigma}^{(1)}$. The transition times are written above the lines, the associated returns, below; transition probabilities are in the spaces.

average cost for the entire "bundle" of trajectories. According to Figure 7, we obtain the expected return for strategy $\boldsymbol{\sigma}^{(1)}$:

$$\begin{aligned} u(\boldsymbol{\sigma}^{(1)}) &= p^2(15+15)+p(1-p)(15+35)+(1-p)p(35+15) \\ &\quad +(1-p)^2(35+35) \\ &= 30p^2+100p(1-p)+70(1-p)^2. \end{aligned}$$

Similarly, we obtain

$$\begin{aligned} u(\boldsymbol{\sigma}^{(2)}) &= 30p^2+105p(1-p)+75(1-p)^2, \\ u(\boldsymbol{\sigma}^{(3)}) &= 30p^2+110p(1-p)+80(1-p)^2, \\ u(\boldsymbol{\sigma}^{(4)}) &= 30p^2+105p(1-p)+75(1-p)^2. \end{aligned}$$

From this, it follows that strategy $\boldsymbol{\sigma}^{(3)}$ is the best one.

The costs $R_{ij}(\boldsymbol{\tau}(i,j);\sigma_i)$ have the following very important property: they are additive in respect to the nonoverlapping parts of the trajectory of the random process under consideration. Additivity is of crucial importance for obtaining a recurrence relation for the total expected cost which, in turn, will be used to find the optimal strategy directly.

From the point of view of constructing mathematical models of preventive maintenance, these costs are sufficiently universal and flexible. In the sequel we demonstrate how through them it is possible to describe many different performance characteristics such as operational readiness, total accumulated losses or expenditures, probability of failure-free operation, etc. Therefore, we shall compare the strategies by comparing the expected costs accumulated during the prescribed number of steps or period of time.† To make such a comparison we need first of all an expression for the expected cost associated with one step in process ξ_t controlled by strategy $\boldsymbol{\sigma}$.

The expected cost associated with the transition from state E_i into E_j is equal, according to the definition of the costs, to the value:

$$\boldsymbol{M}\{R_{ij}(\boldsymbol{\tau}(i,j);\sigma_i\} = r_{ij}(\sigma_i) = \int_0^\infty R_{ij}(x;\sigma_i)\,\mathrm{d}F_{ij}(x;\sigma_i).$$

Averaging all possible transitions from the initial state E_i we obtain an important characteristic called the expected one-step cost:

†For the sake of brevity, the terms "expected cost" or "average cost" will be used instead of "mathematical expectation of the total accumulated cost". The same abbreviation will be used concerning returns.

$$\omega_i(\sigma_i) = \sum_{j=1}^{n} p_{ij}(\sigma_i) r_{ij}(\sigma_i). \tag{3.19}$$

For the sake of brevity, we shall call this value the one-step cost (or the one-step return).

Thus, in the example under consideration

$$\omega_1(\sigma_1 = T_1) = p(kt_1 - c_a) + (1-p)(kT_1 - c_p) = 35 - 20p;$$
$$\omega_1(\sigma_1 = T_2) = p(kt_1 - c_a) + (1-p)(kt_2 - c_a) = 40 - 25p.$$

It will be assumed that the values $\omega_i(\sigma_i)$ are nonnegative for all CAs which are available in state E_i.

Our goal is to find a strategy which minimizes the total expected cost. It is clear that in a more realistic situation, it is not possible to carry out all the tiresome calculations as we did in the above example. It is necessary to have a more efficient mathematical tool to calculate the expected costs. Recurrence relations for the expected accumulated costs serve as such a tool.

A comment seems appropriate here. We are searching for the optimal strategy which will provide the minimal cost or the maximal return. We have already mentioned that we shall restrict ourselves to a consideration only of a special class of strategies called Markovian. Are we depriving ourselves of the possibility of obtaining smaller expected costs by introducing this restriction? Fortunately, the answer is no. In all situations when the controlled process has a finite set of states and finite sets of control actions in every state, the optimal strategies belong to the class of Markovian strategies. This fact is intuitively clear but there is an essential difficulty in its formal proof. We refer the reader who is interested in formal details to the thorough review of Shiryaev[77].

3.4. Finite-step Process

Let us assume that the system is intended for operation only for a fixed number m of steps. The initial state E_i is fixed for the time instant $t_0 = 0$: $\xi_0 = E_i$. The strategy $\boldsymbol{\sigma}$

$$\boldsymbol{\sigma} = \{\sigma_i(0); \sigma_j(k); j = \overline{1, n}; k = \overline{1, m-1}\}, \tag{3.20}$$

is selected. According to above description the CAs are appointed at times $\mathbf{t}_0, \mathbf{t}_1, \ldots, \mathbf{t}_{m-1}$. Denote by $V_i(m, \boldsymbol{\sigma})$ the ME of the total cost accumulated during m steps when the initial state was E_i. This value is the sum of the one-step cost and of the cost accumulated on the subsequent steps. Assume that the first transition was into state E_j ($\xi_{t_1} = E_j$). Denote by $V_j(m-1; \boldsymbol{\sigma}^1, \xi^{(t_1)})$ the ME of the cost accumulated from the moment t_1 up to the end of the operation. This value is calculated under the following conditions:

(a) The strategy on the remaining $m-1$ steps is determined by the "tail" of the strategy $\boldsymbol{\sigma}$ given in (3.20):

$$\boldsymbol{\sigma}^1 = \{\sigma_j(k), j = \overline{1, n}; k = \overline{1, m-1}\}.$$

(b) The transition into state E_j at time $\mathbf{t}_1 = t_1$ is preceded by a definite "history": state E_i, transition time $\mathbf{t}_1$ and the CA $\sigma_i(0)$. All this "history" is denoted by the symbol $\xi^{(t_1)}$. Strictly speaking, we cannot exclude the possibility that this "history" will influence in some way the future evolution of the process; therefore, it has to be reflected in the notation of V_j.

According to the rule of computation, the ME is equal to

$$V_i(m; \boldsymbol{\sigma}) = \sum_{j=1}^{n} p_{ij}(\sigma_i(0)) \times \left\{ \int_0^\infty [R_{ij}(\tau; \sigma_i(0)) + V_j(m-1; \boldsymbol{\sigma}^1; \xi^{(\tau)})] \, dF_{ij}(\tau; \sigma_i(0)) \right\}. \quad (3.21)$$

This process has an important feature: the value $V_j(m-1; \boldsymbol{\sigma}^1; \xi^{(t_1)})$, in fact, does not depend on $\xi^{(t_1)}$:

$$V_j(m-1; \boldsymbol{\sigma}^1; \xi^{(t_1)}) = V_j(m-1; \boldsymbol{\sigma}^1). \quad (3.22)$$

To be convinced of this, it is necessary for the reader to remember how the stochastic mechanism, which determines the evolution of the process after the transition into E_j, works. It depends on the CA $\sigma_j(k)$, $j = \overline{1, n}$, $k = \overline{1, n-1}$. But $\sigma_j(k)$ do not depend on the "history" $\xi^{(t_1)}$ because only Markovian strategies are considered. Later on, the process ξ_t itself has a Markovian property: after the transition into state E_{i_k} the subsequent transitions do not depend on which transitions preceded this one and when they were made. All this reasoning enables us to rewrite (3.21), using (3.19) and (3.22) in the following form:

$$V_i(m, \sigma) = \omega_i(\sigma_i(0)) + \sum_{j=1}^{n} p_{ij}(\sigma_i(0)) V_j(m-1; \boldsymbol{\sigma}^1). \quad (3.23)$$

Now we try to evaluate the method for finding the optimal strategy. According to the statement of the problem, there is a finite set of strategies governing the process on the last $(m-1)$ steps. From this it follows that an optimal strategy exists. Denote

$$\min_{\boldsymbol{\sigma}^1} V_j(m;\boldsymbol{\sigma}^1) = V_j(m;\hat{\boldsymbol{\sigma}}^1) = V_j(m-1), \tag{3.24}$$

that is, the value $V_j(m-1)$ is the minimal expected cost for the initial state E_i by the optimal choice of CAs at times $\mathbf{t}_1, \mathbf{t}_2, \ldots, \mathbf{t}_{m-1}$. The symbol $\hat{\boldsymbol{\sigma}}^1$ is used for the optimal strategy on these last $(m-1)$ steps. Let $\hat{\boldsymbol{\sigma}}$ be the optimal strategy in the m-step process:

$$\min_{\boldsymbol{\sigma}} V_i(m;\boldsymbol{\sigma}) = V_i(m;\hat{\boldsymbol{\sigma}}) = V_i(m). \tag{3.25}$$

Without loss in generality it may be assumed that there is only one such optimal strategy.

The fact of crucial importance is that the CAs prescribed by the optimal strategy $\hat{\boldsymbol{\sigma}}$ for the time instants $\mathbf{t}_1, \mathbf{t}_2, \ldots, \mathbf{t}_{m-1}$ *coincide* with the CAs prescribed by the strategy $\hat{\boldsymbol{\sigma}}^1$ for the same time instants. This property can be proved directly and it is equivalent to Bellman's dynamic programming principle (see [18, p. 86]):

> "An optimal policy has the property that whatever the initial state and initial decision are, the remaining decisions must constitute an optimal policy with regard to the state resulting from the first decision."

Assume that $\hat{\boldsymbol{\sigma}}^1$ and $V_j(m-1)$ are known. To obtain $\hat{\boldsymbol{\sigma}}$ we must find only the optimal $\sigma_i(0)$. Let us consider all possible strategies obtained by combining the CA $\sigma_i(0)$ and $\hat{\boldsymbol{\sigma}}^1$ and choose the best one; this will give us the optimal strategy $\hat{\boldsymbol{\sigma}}$. Thus, we have

$$\min_{\boldsymbol{\sigma}} V_i(m;\boldsymbol{\sigma}) = V_i(m;\hat{\boldsymbol{\sigma}}) = V_i(m)$$

$$= \min_{\sigma_i(0)} \left\{ \omega_i(\sigma_i(0)) + \sum_{j=1}^{n} p_{ij}(\sigma_i(0)) V_j(m-1) \right\}. \tag{3.26}$$

There now remains to demonstrate how to use that recurrence relation for obtaining the optimal strategy in an explicit form. (If the reader is familiar with the standard "backward" dynamic programming computation technique, he can ignore the following few paragraphs.)

Assume that only one step is left and the optimal CA at time $\mathbf{t}_{m-1}$ for state E_i has to be selected. It is clear that we have to use the following formula:

$$\min_{\sigma_i(m-1)\in S_i} \omega_i(\sigma_i(m-1)) = \omega_i(\hat{\sigma}_i^*(m-1)) = V_i(1), \qquad i = \overline{1, n}.$$

Knowing $V_i(1)$ for $i = 1, \ldots, n$, we shall start the "backward" motion. Let us write relation (3.26) for a two-step process starting at the moment $\mathbf{t}_{m-2}$:

$$V_i(2) = \min_{\sigma_i(m-2)\in S_i} \left\{\omega_i(\sigma_i(m-2)) + \sum_{j=1}^{n} p_{ij}(\sigma_i(m-2))V_j(1)\right\}$$

$$= \omega_i(\hat{\sigma}_i^*(m-2)) + \sum_{j=1}^{n} p_{ij}(\sigma_i^*(m-2))V_j(1).$$

The values $V_j(3)$ will be found in a similar way by the use of $V_j(2)$, and so forth. In other words, the following recurrence system is used for obtaining the optimal strategy:

$$V_i(k) = \min_{\sigma_i(m-k)\in S_i} \left\{\omega_i(\sigma_i(m-k)) + \sum_{j=1}^{n} p_{ij}(\sigma_i(m-k))V_j(k-1)\right\},$$

$$i = 1, \ldots, n; \quad k = 1, \ldots, m; \quad V_j(0) \equiv 0. \qquad (3.27)$$

If $\hat{\sigma}_i^*(k)$, $i = \overline{1, n}$, $k = m, m-1, \ldots$, are the CAs which provide the minimal values for the right side in relations (3.27), then the strategy $\hat{\boldsymbol{\sigma}}^* = \{\hat{\sigma}_i^*(l); i = \overline{1, n}; l = \overline{0, m-1}\}$ is the optimal one in a m-step process.

We note that strategy $\hat{\boldsymbol{\sigma}}$ in the finite-step process in the general case is the second-type strategy (see Section 3.2, B).

The system (3.27) can be written in a compact form using vector and matrix notations. Let us define the following one-column matrices (vectors):

$$\mathbf{V}(k) = \begin{Vmatrix} \dot{V}_1(k) \\ \cdots \\ \cdots \\ \cdots \\ V_n(k) \end{Vmatrix}, \qquad \boldsymbol{\Omega}(\sigma(m-k)) = \begin{Vmatrix} \omega_1(\sigma_1(m-k)) \\ \cdots\cdots \\ \cdots\cdots \\ \cdots\cdots \\ \omega_n(\sigma_n(m-k)) \end{Vmatrix}.$$

The system (3.27) can be written in the form:

$$\mathbf{V}(k) = \min_{\sigma(m-k)} \{\boldsymbol{\Omega}(\sigma(m-k)) + \mathbf{P}(\sigma(m-k))\mathbf{V}(k-1)\},$$

$$k = 1, \ldots, m. \qquad (3.28)$$

If the problem is stated so as to maximize the return, all these relations remain valid, and only the symbol *max* instead of *min* should be used. Then, of course, the values $V_j(m)$ will give the *maximal* returns in a m-step process.

Perhaps the reader has obtained the impression that the costs or returns are adapted only to description of such purely "economic" characteristics as expenses of running the equipment or the profit gained from it. We give a simple example to demonstrate that the costs can represent a probabilistic characteristic of the system.

Assume a subset I_F of states considered as failure states. After entering state $E_i \in I_F$, the system stops working and that event is considered as a failure. We shall show by a special definition of one-step costs how it is possible to express through the values $V_j(m)$ the probability of a failure-free operation during m steps. Let $\bar{I}_F$ denote the complement set of I_F, and let R_{ij} be penalties associated with transitions:

$$R_{ij} = \begin{cases} 1, & \text{if } E_i \in \bar{I}_F,\ E_j \in I_F; \\ 0, & \text{if } E_i \in \bar{I}_F,\ E_j \in \bar{I}_F \text{ or } E_i \in I_F. \end{cases}$$

The formula for the minimal one-step expected cost now is ($E_i \in \bar{I}_F$):

$$V_i(1) = \min_{\sigma_i(m-1) \in S_i} \left\{ \sum_{j \in I_F} p_{ij}(\sigma_i(m-1)) \right\}.$$

It is easy to check that $V_i(1)$ is equal to the minimal probability of the failure in a one-step process. The expression for $V_i(2)$ is the following:

$$V_i(2) = \min_{\sigma_i(m-2) \in S_i} \left\{ \sum_{j\,:\,E_j \in I_F} p_{ij}(\sigma_i(m-2)) + \sum_{j\,:\,E_j \in \bar{I}_F} p_{ij}(\sigma_i(m-2)) V_j(1) \right\}.$$

The value in the braces is the probability that the system does fail in the two-step process. The failure occurs either on the first step or, if the transition has been made into state $E_i \in \bar{I}_F$, on the second step. By induction this proof can be given for an arbitrary number of steps.

3.5. *Finite-time Process*

Assume that the system must operate only a fixed period of time $[0, \theta]$. The counting of costs (returns) also will be done up to time instant $t = \theta$. A special supposition must be made with respect to the cost associated with the step which will remain unfinished up to time θ. Assume that in the process ξ_t, the transitions occurred at the instants $t_1, t_2, \ldots, t_r, t_{r+1}$, and $t_r < \theta < t_{r+1}$, that is the rth step remained unfinished. Assume that if at instant t_r the transition is made into state E_i, the CA σ_i is selected and E_i is headed by E_j, the cost $R_{ij}(\theta - t_r; \sigma_i)$ will be associated with the last, incomplete step. Figure 8 shows the accumulation of costs for one sample function of the process ξ_t. In the case pictured in Figure 8 the total cost

$$R_{12}(t_1; \sigma_1) + R_{2n}(t_2 - t_1; \sigma_2) + R_{ni}(t_3 - t_2; \sigma_n) + R_{ij}(\theta - t_3; \sigma_i)$$

will be accumulated.

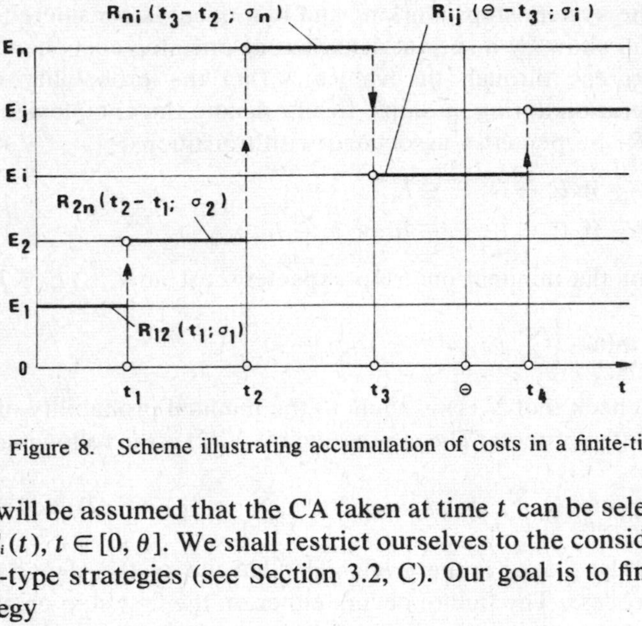

Figure 8. Scheme illustrating accumulation of costs in a finite-time process.

It will be assumed that the CA taken at time t can be selected from the set $S_i(t)$, $t \in [0, \theta]$. We shall restrict ourselves to the consideration of the third-type strategies (see Section 3.2, C). Our goal is to find the optimal strategy

$$\boldsymbol{\sigma} = \hat{\boldsymbol{\sigma}}(t) = \{\hat{\sigma}_1(t), \hat{\sigma}_2(t), \ldots, \hat{\sigma}_n(t)\}, \tag{3.29}$$

providing the minimal expected cost accumulated in the time interval $[0, \theta]$.

The state E_i is assumed to be the initial one at the time $t = 0$: $\xi_0 = E_i$. Assume that the optimal strategy $\hat{\boldsymbol{\sigma}}(t)$ does exist and the corresponding minimal expected cost is equal to $v_i(\theta)$.

Let the CA selected at $t_0 = 0$ be $\sigma_i \in S_i(0)$ and the first transition occur at time τ into state E_j. The same optimal strategy $\hat{\boldsymbol{\sigma}}(t)$ on the time interval $[\tau, \theta]$ will be preserved. Denote by $v_j(\theta - \tau)$ the ME of the total accumulated cost on the interval $[\tau, \theta]$ when the initial state at $t = \tau$ is E_j. By using the same arguments as in Section 3.4, it may be shown that the value $v_j(\theta-\tau)$ does not depend on the "history" $\xi^{(\tau)}$ before time τ. It may be stated that the use of strategy $\hat{\boldsymbol{\sigma}}(t)$ during the interval $[\tau, \theta]$ provides the minimal expected cost accumulated on that interval. Now let us establish the relations between $v_j(\theta)$ and $v_j(\theta-\tau)$.

The first step with the probability

$$p_\tau(j) = p_{ij}(\sigma_i)[F_{ij}(\tau+\Delta\tau;\sigma_i) - F_{ij}(\tau;\sigma_i)]$$

will be finished with the transition into E_j occurring in the time interval $(\tau, \tau+\Delta\tau)$, $j = 1, 2, \ldots, n$. $q_j = [1-F_{ij}(\theta;\sigma_i)]p_{ij}(\sigma_i)$ is the probability that the transition into E_j will occur after $t=\theta$. $p_\tau(j)$ is the probability that cost $R_{ij}(\tau;\sigma_i)$ will be accumulated during the first step plus $v_j(\theta-\tau)$ during the remaining period of time. q_j is the probability that the only cost is $R_{ij}(\theta;\sigma_i)$, $j = 1, \ldots, n$. Therefore, the expected cost accumulated during the period $[0, \theta]$ with the initial state E_i and by the use of the strategy $\hat{\boldsymbol{\sigma}}(t)$ is equal to

$$\sum_{j=1}^{n}\int_0^\theta p_{ij}(\sigma_i)[R_{ij}(\tau;\sigma_i) + v_j(t-\tau)]\,\mathrm{d}F_{ij}(\tau;\sigma_i) + \sum_{j=1}^{n} p_{ij}(\sigma_i) \times [1-F_{ij}(\theta;\sigma_i)]R_{ij}(\theta;\sigma_i).$$

Now there remains only to select the optimal $\sigma_i \in S_i(0)$ so as to minimize the total expected cost. Then we obtain the desired recurrence relations:

$$v_i(\theta) = \min_{\sigma_i\in S_i(0)}\left\{\sum_{j=1}^{n} p_{ij}(\sigma_i)\int_0^\theta R_{ij}(\tau;\sigma_i)\,\mathrm{d}F_{ij}(\tau;\sigma_i) + \sum_{j=1}^{n} p_{ij}(\sigma_i)[1-F_{ij}(\theta;\sigma_i)]R_{ij}(\theta;\sigma_i) + \sum_{j=1}^{n} p_{ij}(\sigma_i)\int_0^\theta v_j(\theta-\tau)\,\mathrm{d}F_{ij}(\tau;\sigma_i)\right\}, \quad i = 1, \ldots, n. \tag{3.30}$$

Unfortunately, we cannot use them for finding the optimal strategy directly.

We now describe the method for obtaining the approximate solution; it is based on the use of discrete time scale.

Assume that all DFs $F_{ij}(\tau;\sigma_i)$ are stepwise and "jumps" occur only at points which are multiples of some constant Δ, that is, on points $t = \Delta, 2\Delta, \ldots$. If we select a sufficiently small Δ, we can approximate every continuous distribution to the desired degree. Assume also that period θ is a multiple of $\Delta: \theta = N\Delta$. In that case integrals in (3.30) may be replaced by finite sums. If we denote

$$F_{ij}(r\Delta;\sigma_i) - F_{ij}((r-1)\Delta;\sigma_i) = f_{ij}(r\Delta;\sigma_i),\ r \geqslant 1,$$

$$f_{ij}(0;\sigma_i) = 0,$$

we obtain the following form for relations (3.30):

$$v_i(N\Delta) = \min_{\sigma_i \in S_i(0)} \left\{ \sum_{j=1}^{n} p_{ij}(\sigma_i) \sum_{r=0}^{N} R_{ij}(r\Delta;\sigma_i) f_{ij}(r\Delta;\sigma_i) + \sum_{j=1}^{n} p_{ij}(\sigma_i)[1 - F_{ij}(N\Delta;\sigma_i)] R_{ij}(N\Delta;\sigma_i) + \sum_{j=1}^{n} p_{ij}(\sigma_i) \sum_{r=0}^{N} v_j(N\Delta - r\Delta) f_{ij}(r\Delta;\sigma_i) \right\}. \tag{3.31}$$

These relations may be used directly for computation by applying the previously described "backward motion" technique. Assume that there remains only a time span Δ until the end of the operation, that is, the initial is the time instant $t = (N-1)\Delta$. Then, setting $N = 1$ in (3.31) we obtain:

$$v_i(\Delta) = \min_{\sigma_i \in S_i((N-1)\Delta)} \left\{ \sum_{j=1}^{n} p_{ij}(\sigma_i) R_{ij}(\Delta;\sigma_i) f_{ij}(\Delta;\sigma_i) + \sum_{j=1}^{n} p_{ij}(\sigma_i)[1 - F_{ij}(\Delta;\sigma_i)] R_{ij}(\Delta;\sigma_i) \right\}, \quad i = 1, \ldots, n. \tag{3.32}$$

This relation serves for finding optimal CA $\sigma_i((N-1)\Delta)$, $i = 1, 2, \ldots, n$. Similarly, for the initial time $t = (N-2)\Delta$ we have

$$v_i(2\Delta) = \min_{\sigma_i \in S_i((N-2)\Delta)} \left\{ \sum_{j=1}^{n} p_{ij}(\sigma_i) \sum_{r=0}^{2} R_{ij}(r\Delta;\sigma_i) f_{ij}(r\Delta;\sigma_i) + \sum_{j=1}^{n} p_{ij}(\sigma_i)[1 - F_{ij}(2\Delta;\sigma_i)] R_{ij}(2\Delta;\sigma_i) + \sum_{j=1}^{n} p_{ij}(\sigma_i) \sum_{r=0}^{2} v_j(2\Delta - r\Delta) f_{ij}(r\Delta;\sigma_i) \right\}. \tag{3.33}$$

The right side of these relations does not contain unknown values because $v_j(2\Delta) f_{ij}(0;\sigma_i) = 0$ and $v_j(\Delta)$ were found on the previous step. The computation proceeds in a similar way for $v_i(k\Delta)$, $k \geqslant 3$.

In Section 1 of Chapter 2 we give a numerical example in order to demonstrate the use of these discrete time recurrence relations for finding an optimal strategy.

3.6. The Process with an Absorbing State

In several models, we shall deal with a process ξ_t which has at least one absorbing state, say E_n. That means that the system can move from every state E_i, $i \neq n$, to E_n but after entering state E_n remains there forever. In many cases such a "trapping" state corresponds to the failure of the system. Sometimes the process ξ_t itself does not have an absorbing state but, according to the statement of the problem, the functioning of the system might be stopped after each transition with nonzero probability β, $0 < \beta < 1$. Formally, this is equivalent to a presence of a fictitious absorbing state E_{n+1}, which the process can enter on every transition with probability β.

Considering the question of finding the optimal strategy, we shall not restrict ourselves to some prescribed time interval or number of steps. The process is observed and, therefore, the costs are accumulated up to the moment it enters the absorbing state.

It will be assumed that one and the same state E_n remains an absorbing one under all possible strategies and that it is achievable from every other state.

Let us take an arbitrary stationary Markovian strategy $\boldsymbol{\sigma} = (\sigma_1, \sigma_2, \ldots, \sigma_{n-1})$, $\sigma_i \in S_i$. It was proved (see, for example, [7], [26], [28]) that the optimal strategy belongs to the class S of all Markovian stationary strategies of the first type.

Let the system start to work at $t = 0$ with the initial state E_i; a certain strategy $\boldsymbol{\sigma}$ is selected. Denote by V_i^{σ} the expected total cost accumulated from time $t = 0$ up to the entrance into the absorbing state. One can prove that V_i^{σ} is a finite value for every i. Intuitively it is obvious because all sample functions of the process with probability 1 will enter the absorbing state during a finite time. Let $\hat{\boldsymbol{\sigma}}$ be the optimal strategy:

$$\min_{\sigma \in S} V_i^{\sigma} = V_i^{\hat{\sigma}} = V_i.$$

Suppose that process ξ_t, governed by strategy $\hat{\boldsymbol{\sigma}}$, went at instant t into state E_j. Denote by $V_j(t\,;\infty)$ the expected cost accumulated on the interval (t, ∞), i.e. from instant t up to the visit into the absorbing state. The behaviour of the process ξ_t on this interval in a statistical sense is the same as if the process had started initially from state E_j at $t=0$. This important property is provided by the Markovian character of the strategy governing the process and the nature of the process itself. Therefore, $V_j^{\sigma}(t;\infty)=V_j$. Now we are ready to write the recurrence relations for V_j.

The total cost is the sum of the cost on the first step and the "future" cost. Assume that a CA σ_i is selected initially for state E_i and after the first transition the process is controlled optimally by strategy $\hat{\boldsymbol{\sigma}}$. Then we obtain the recurrence relation for the total expected cost V_i^*:

$$V_i^* = \omega_i(\sigma_i) + \sum_{j=1}^{n} p_{ij}(\sigma_i)V_j.$$

The choice of the best CA at time $t=0$ leads us to the desired relation:

$$V_i = \min_{\sigma_i \in S_i} \left[\omega_i(\sigma_i) + \sum_{j=1}^{n} p_{ij}(\sigma_i)V_j \right], \qquad i = 1, \ldots, n-1, \tag{3.34}$$

which can be rewritten in the matrix notations as follows:

$$\mathbf{V} = \min_{\boldsymbol{\sigma}} [\boldsymbol{\Omega}(\boldsymbol{\sigma}) + \mathbf{P}_1(\boldsymbol{\sigma})\mathbf{V}], \tag{3.35}$$

where

$$\mathbf{P}_1(\boldsymbol{\sigma}) = \|p_{ij}(\sigma_i)\|, \qquad i, j = 1, 2, \ldots, n-1.$$

We describe an algorithm for obtaining the near-optimal values V_j and the near-optimal strategy. It will be assumed that $\omega_i(\sigma_i) \geq C > 0$ for all σ_i, $i = 1, \ldots, n-1$.

Step 1. Select an arbitrary stationary strategy $\boldsymbol{\sigma}^{(0)}$ and solve the set of equations

$$\hat{\mathbf{V}}^{(0)} = \boldsymbol{\Omega}(\boldsymbol{\sigma}^{(0)}) + \mathbf{P}_1(\boldsymbol{\sigma}^{(0)})\hat{\mathbf{V}}^{(0)}. \tag{3.36}$$

Step 2. Construct the sequence $\hat{\mathbf{V}}^{(k)}$ according to the following rule:

$$\hat{\mathbf{V}}^{(k)} = \min_{\boldsymbol{\sigma}} [\boldsymbol{\Omega}(\boldsymbol{\sigma}) + \mathbf{P}_1(\boldsymbol{\sigma})\hat{\mathbf{V}}^{(k-1)}], \qquad k = 1, 2, \ldots. \tag{3.37}$$

It should be noted that this sequence decreases monotonically and converges to the vector of minimal expected costs. The detailed proof is given in [35].

Step 3. Construct the sequence $\mathbf{V}^{(l)}$ according to the rule:

$$\mathbf{V}^{(l)} = \min_{\boldsymbol{\sigma}} [\boldsymbol{\Omega}(\boldsymbol{\sigma}) + \mathbf{P}_1(\sigma)\mathbf{V}^{(l-1)}], \qquad l = 1, 2, \ldots, \tag{3.38}$$

where $\mathbf{V}^{(0)}$ is the zero-vector. It can be proved that this sequence monotonically increases and also tends to the optimal vector $\mathbf{V}$.

Assume that we have obtained such values $\hat{\mathbf{V}}^{(k)}$ and $\mathbf{V}^{(l)}$ that for $\epsilon > 0$

$$|\hat{\mathbf{V}}^{(k)} - \mathbf{V}^{(l)}| < \epsilon$$

and let

$$\hat{\mathbf{V}}^{(k+1)} = \min_{\boldsymbol{\sigma}} [\boldsymbol{\Omega}(\boldsymbol{\sigma}) + \mathbf{P}_1(\boldsymbol{\sigma})\hat{\mathbf{V}}^{(k)}] = \boldsymbol{\Omega}(\boldsymbol{\sigma}^{(k)}) + \mathbf{P}_1(\boldsymbol{\sigma}^{(k)})\hat{\mathbf{V}}^{(k)}.$$

Then the strategy $\boldsymbol{\sigma}^{(k)}$ may be assumed as an ϵ-optimal one: using this strategy we obtain the values of total expected cost which differ from the optimal vector not more than for the value ϵ.

Now we discuss one way of counting payoffs often used in economic calculations. The case in question is termed as *discounting*. Assume that in the calculation is involved some amount of money (cost, return) $V(\tau)$ which has to be paid after a time τ. A positive constant α is chosen and the mentioned amount is reduced to the present time by multiplying its value by a constant $e^{-\alpha\tau}$. In other words, the expenses which will appear in the future in comparison with the expenses paid now must be taken into account with some reduction. That is a natural consequence of the existence of the discount rate. We shall not go into details; the reader is referred, for example, to the book of Jardine [49]. We are interested only in the formal interpretation of discounting.

Let a certain first-type strategy be selected. The rules for counting payoffs lead to the following relation:

$$v_i(\alpha) = \omega_i(\alpha) + \sum_{j=1}^{n} p_{ij}(\sigma_i) \int_0^\infty e^{-\alpha t} v_j(\alpha)\, dF_{ij}(t;\sigma_i);$$

$$\omega_i(\alpha) = \sum_{j=1}^{n} p_{ij}(\sigma_i) \int_0^\infty \left\{ \int_0^t e^{-\alpha x}\, dR_{ij}(x;\sigma_i) \right\} dF_{ij}(t;\sigma_i),$$

where $v_i(\alpha)$ are the total payoffs in the infinite-time process, given the initial state E_i and the discount factor $\alpha > 0$ (see [50]). Denote

$$\tilde{p}_{ij}(\sigma_i, \alpha) = \int_0^\infty e^{-\alpha t} p_{ij}(\sigma_i)\, dF_{ij}(t;\sigma_i).$$

Now the formula for $v_i(\alpha)$ can be rewritten in the following form:

$$v_i(\alpha) = \omega_i(\alpha) + \sum_{j=1}^{n} \tilde{p}_{ij}(\sigma_i, \alpha) v_j(\alpha).$$

We can consider this relation as a recurrence one for the expected payoffs in an infinite-step process which has nonzero "stopping" probability at every step. For state E_i this probability is equal to the value $[1 - \Sigma_{j=1}^{n} \tilde{p}_{ij}(\sigma_i, \alpha)]$.

3.7. Process on the Infinite Time Interval

Now a random process ξ_t which is observed during the infinite period of time will be discussed. Although there are no technical systems whose life span is infinitely large, the infinite-time model serves as a good formal description for long-term operating systems whose operation time is long enough in comparison with the length of one working cycle.

The characteristic feature of this case is that for an infinitely long operation time the costs may be also infinitely large. We have mentioned above that in such a situation the suitable criterion would be the average cost rate (expected cost per unit time). The complication which arises here is that the cost rate is a nonadditive function in respect to the time or number of steps in the random process. That compels us to use a different mathematical technique not considered before in our exposition.

Let us describe all necessary prerequisites. Only Markovian stationary strategies will be considered. There is a finite number of CAs in every set S_i, $i = 1, \dots, n$. For every strategy $\boldsymbol{\sigma} = (\sigma_1, \dots, \sigma_n)$, the matrix of transition probabilities $\mathbf{P} = \|p_{ij}(\sigma_i)\|$ defines a Markov chain having all states communicating: for every two different states E_i and E_j there is a way to go from E_i to E_j. (The reader is referred to the monograph of Feller [32, Chapter 15] for the most complete description of Markov chain properties.)

Let us go back to the recurrence relation (3.30) written for the finite time. We agree upon the use of a strategy $\boldsymbol{\sigma}$ of the first type for all values of t: $\boldsymbol{\sigma}(t) \equiv \boldsymbol{\sigma} = (\sigma_i, i = 1, \dots, n)$. Let $v_i(\theta, \boldsymbol{\sigma})$ denote the total expected cost on the interval $[0, \theta]$ calculated under the condition that the initial state was E_i and the strategy $\boldsymbol{\sigma}$ was used constantly. Then the expression for $v_i(\theta; \boldsymbol{\sigma})$ is:

$$v_i(\theta; \boldsymbol{\sigma}) = \omega_i(\theta; \sigma_i) + \Psi_i(\theta; \sigma_i) + \sum_{j=1}^{n} p_{ij}(\sigma_i) \int_0^{\theta} v_j(\theta - \tau; \boldsymbol{\sigma}) \, \mathrm{d}F_{ij}(\tau; \sigma_i),$$

$$i = 1, \dots, n, \qquad (3.39)$$

where we denote

$$\omega_i(\theta;\sigma_i)=\sum_{j=1}^{n}p_{ij}(\sigma_i)\int_0^{\theta}R_{ij}(\tau;\sigma_i)\,\mathrm{d}F_{ij}(\tau;\sigma_i),$$

$$\Psi_i(\theta;\sigma_i)=\sum_{j=1}^{n}p_{ij}(\sigma_i)[1-F_{ij}(\theta;\sigma_i)]R_{ij}(\theta;\sigma_i). \tag{3.40}$$

It should be noted that

$$\lim_{\theta\to\infty}\omega_i(\theta;\sigma_i)=\omega_i(\sigma_i)=\sum_{j=1}^{n}p_{ij}(\sigma_i)r_{ij}(\sigma_i)$$

$$\lim_{\theta\to\infty}\Psi_i(\theta;\sigma_i)=0. \tag{3.41}$$

In the sequel, we need the formula for the expected holding time ν_i in state E_i or more precisely, the ME of the length of one step starting from state E_i. Averaging the expected transition times from E_i into E_j over all possible transitions, we obtain

$$\nu_i(\sigma_i)=\sum_{j=1}^{n}p_{ij}(\sigma_i)M\{\boldsymbol{\tau}(i,j)\}=\sum_{j=1}^{n}p_{ij}(\sigma_i)\int_0^{\infty}\tau\ \mathrm{d}F_{ij}(\tau;\sigma_i). \tag{3.42}$$

It will be assumed that all $\nu_i(\sigma_i)$ are finite.

The fact which plays a very important role in further exposition is that values $v_i(\theta;\boldsymbol{\sigma})$ have the following asymptotic behaviour:

$$v_i(\theta;\boldsymbol{\sigma})\underset{\theta\to\infty}{=}g(\boldsymbol{\sigma})\theta+w_i(\boldsymbol{\sigma})+o(1), \tag{3.43}$$

where

$$g(\boldsymbol{\sigma})=\frac{\sum_{i=1}^{n}\pi_i(\boldsymbol{\sigma})\omega_i(\sigma_i)}{\sum_{i=1}^{n}\pi_i(\boldsymbol{\sigma})\nu_i(\sigma_i)}, \tag{3.44}$$

and $w_i(\boldsymbol{\sigma})$ are constants depending on p_{ij} and F_{ij}; $\omega_i(\sigma_i)$ and $\nu_i(\sigma_i)$ were defined by (3.42) and (3.41); $\pi_i(\boldsymbol{\sigma})$, $i=1,2,\ldots,n$, is a solution of the set of equations

$$\boldsymbol{\pi}=\boldsymbol{\pi}\mathbf{P},$$

$$\sum_{i=1}^{n}\pi_i=1, \tag{3.45}$$

$$\boldsymbol{\pi}=(\pi_1,\ldots,\pi_n).$$

The factor $g(\sigma)$ in (3.43) determines the expected cost (or return) per unit time over an infinite period of the functioning of the system. It will be termed "cost (or return) rate" or "expected cost (or return) per unit time".

For the validity of (3.43) some additional conditions must be held. In practical situations this does not lead to any complications; this question is discussed in Remark 2 of Appendix 2.

Let us now discuss the formal and physical meaning of the values π_i. It is well-known from the theory of Markov chains that in the case of all communicating states, the system of equations (3.45) always has a solution which is unique and nonnegative. If we denote l_{mm} the ME of the number of steps of the first return time for state E_m, then from the Markov chains theory it follows that $\pi_m = 1/l_{mm}$. Therefore, the values π_m are equal to the average return rate into state E_m. Let us assume now that a large number K of transitions has been made in the SMP ξ_t. On the average, once during every K/l_{mm} steps, the process visits state E_m. Therefore, during K steps there were approximately K_m visits into state E_m, where $K_m = K/l_{mm} = \pi_m K$.

Therefore, π_m permits an interpretation as a rate of steps starting from E_m.

Formula (3.43) means that for all states the expected costs are asymptotically proportional to the time variable θ with the *same* factor of proportionality $g(\sigma)$.

The complete evaluation of the expression for $g(\sigma)$ is comparatively complicated (see [50]). We shall omit the formal computation and instead of that we shall try to clarify further the physical meaning of formula (3.44).

Let θ be large. During this time, K_i transitions occurred into E_i, $i = 1, \ldots, n$. After each such transition the process stayed in this state, on an average, during time ν_i (we omit the symbol denoting the strategy σ). Therefore θ is approximately equal to the following sum:

$$\theta \approx K_1\nu_1 + K_2\nu_2 + \cdots + K_n\nu_n.$$

On the other hand, every visit into state E_i gives a cost whose expected value is ω_i. The total expected cost accumulated for the time θ is equal to the value

$$v_i(\theta) \approx \sum_{i=1}^{n} K_i\omega_i.$$

If we now write the expression for the total expected cost per unit time, then we obtain the following formula:

$$g=\frac{v_i(\theta)}{\theta}\approx\Big(\sum_{i=1}^{n}K_i\omega_i\Big)\Big/\Big(\sum_{i=1}^{n}K_i\nu_i\Big)$$

$$=\left(\sum_{i=1}^{n}\frac{K_i}{\sum_{i=1}^{n}K_i}\omega_i\right)\Big/\left(\sum_{i=1}^{n}\frac{K_i}{\sum_{i=1}^{n}K_i}\nu_i\right).$$

Note that $K_i/\Sigma_{j=1}^{n}K_j$ is equal to the average fraction of steps which the system has spent in state E_i. For a large θ, this fraction is equal approximately to π_i. Thus we obtained the expression (3.44).

The solution of (3.45), $\boldsymbol{\pi}=(\pi_1,\pi_2,\ldots,\pi_n)$ is termed a stationary or limiting distribution. We shall use "stationary probabilities" for denoting π_i.

It was pointed out before that for a long-term operating system the probability of being in an operable state at an arbitrary moment θ, $\theta\to\infty$, is an important performance characteristic. We demonstrate that this characteristic can be expressed in terms of $g(\boldsymbol{\sigma})$ by an appropriate choice of one-step costs. Let $\mathbf{E}_J=\{E_i, i\in J\}$ denote the set of all operable states of our system. Let

$$R_{ij}(\boldsymbol{\tau};\sigma_i)=\begin{cases}\boldsymbol{\tau}, & i\in\mathbf{J},\\ 0, & i\notin\mathbf{J},\end{cases}$$

or, in other words, the cost (more pertinent here would be the term "return") is equal to the transition time from state E_i into E_j. Then we obtain

$$\omega_i(\sigma_i)=\begin{cases}\nu_i(\sigma_i), & i\in\mathbf{J},\\ 0, & i\notin\mathbf{J},\end{cases}$$

and formula (3.44) takes the following form:

$$g(\boldsymbol{\sigma})=\Big(\sum_{i\in J}\pi_i(\boldsymbol{\sigma})\nu_i(\sigma_i)\Big)\Big/\Big(\sum_{i=1}^{n}\pi_i(\boldsymbol{\sigma})\nu_i(\sigma_i)\Big). \tag{3.46}$$

It is intuitively clear now that $g(\boldsymbol{\sigma})$ expresses the rate of time which the system spends in nonfailure states. Therefore, $g(\boldsymbol{\sigma})=R_\infty(0)$, where $R_\infty(0)$ is the coefficient of operational readiness (see 2.4). Appendix 2 contains the proof of the fact that (3.46) is equal to the stationary probability that the process ξ_t is in the set $\mathbf{E}_J$ when $t\to\infty$.

Our goal is to obtain a strategy $\boldsymbol{\sigma}$ which minimizes the value $g(\boldsymbol{\sigma})$. If the total number of strategies is not large, that can be done by means of enumeration, by the direct use of expression (3.44). If the number of strategies is very large, an alternative method seems much more convenient. This method is embodied in the following algorithm proposed by Howard [4], [46], [48, Vol. 2, Chapter 15].

Let us substitute (3.43) in the recurrence relation (3.39):

$$g(\boldsymbol{\sigma})\theta + w_i(\boldsymbol{\sigma}) + o(1) = \omega_i(\theta; \sigma_i) + \Psi_i(\theta; \sigma_i) + \sum_{j=1}^{n} p_{ij}(\sigma_i)\left\{\int_0^\theta g(\boldsymbol{\sigma})(\theta - \tau)\,\mathrm{d}F_{ij}(\tau; \sigma_i) + \int_0^\theta w_i(\boldsymbol{\sigma})\,\mathrm{d}F_{ij}(\tau; \sigma_i)\right\}. \tag{3.47}$$

After some manipulations we obtain, taking the limit $\theta \to \infty$ and using (3.41), the following set of equations:

$$w_i(\boldsymbol{\sigma}) + g(\boldsymbol{\sigma})\nu_i(\sigma_i) = \omega_i(\sigma_i) + \sum_{j=1}^{n} p_{ij}(\sigma_i) w_j(\boldsymbol{\sigma}), \qquad i = 1, 2, \ldots, n, \tag{3.48}$$

In this set there are n equations and $(n+1)$ variables: w_i, $i = 1, \ldots, n$, and $g(\boldsymbol{\sigma})$. It is possible to demonstrate that this system has a unique solution up to an arbitrary additive constant. Setting $w_n(\boldsymbol{\sigma}) = 0$ we obtain a system with the number of equations equal to the number of variables. The value $g(\boldsymbol{\sigma})$ corresponding to the optimal strategy may be found in the following manner.

Step 1. Select an arbitrary strategy $\boldsymbol{\sigma} = (\sigma_1, \sigma_2, \ldots, \sigma_n)$ and find the values $\omega_i(\sigma_i)$, $p_{ij}(\sigma_i)$, $\nu_i(\sigma_i)$.

Step 2. Setting $w_n(\boldsymbol{\sigma}) = 0$, solve the system of equations (3.48) in respect to the cost rate g and variables w_i, $i = 1, \ldots, n$.

Step 3. For each state E_i find the CA σ_i^* which *minimizes* the ratio

$$z_i(\sigma_i) = \frac{\omega_i(\sigma_i) + \sum_{j=1}^{n} p_{ij}(\sigma_i) w_j - w_i}{\nu_i(\sigma_i)}, \tag{3.49}$$

using the present values of w_i.

Make σ_i^* a new CA for the ith state. If the strategy is identical with the one from the last cycle, then $\boldsymbol{\sigma}^* = (\sigma_1, \ldots, \sigma_n)$ is the optimal strategy; otherwise start Step 1 with the strategy $\boldsymbol{\sigma}_1 = \boldsymbol{\sigma}^*$.

Remarks. It was mentioned above that the Markov chain imbedded in the SMP must have all states interconnectible. This restriction can be weakened.

Assume that there is a strategy which generates a Markov chain having only one closed subset $\mathbf{E}_1 \subset \mathbf{E}$ of interconnecting states such, that from every state $E_j \in \mathbf{E} - \mathbf{E}_1$, it is possible to get into set $\mathbf{E}_1$. From the definition of $\mathbf{E}_1$, it follows that it is impossible to return back into set $\mathbf{E} - \mathbf{E}_1$ from $E_i \in \mathbf{E}_1$. The set $\mathbf{E}_1$ plays the role of a generalized "trapping" set. The states belonging to the set $\mathbf{E} - \mathbf{E}_1$ will be referred to as transient. The evolution of the process starting in a transient state E_j consists of a number of steps "inside" set $\mathbf{E} - \mathbf{E}_1$, the transition into one of the states belonging to $\mathbf{E}_1$ and an infinite number of transitions "inside" the set $\mathbf{E}_1$. It is possible to demonstrate that the expected accumulated cost $v_j(\theta)$ has the asymptotic behaviour according to formula (3.43). We illustrate this statement with a heuristic reasoning.

Assume θ to be large and consider a trajectory of ξ_t starting at $t = 0$ in the transient state E_i. After some walking time τ^* which is short in comparison with θ, the process will enter set $\mathbf{E}_1$. Assume that first visit into $\mathbf{E}_1$ took place into state E_j and the probability of that event is equal to p^*_{ij}. Note that

$$\sum_{E_j \in \mathbf{E}_1} p^*_{ij} = 1.$$

Then

$$v_i(\theta) \approx \sum_{E_j \in \mathbf{E}_1} p^*_{ij}(q_{ij} + v_j(\theta - \tau^*)),$$

where q_{ij} is the expected cost accumulated during the walk in the transient states. From this formula and from (3.43) it is clear that, taking $\theta \to \infty$, we obtain $v_i(\theta) \approx g \cdot \theta + w^*_j$.

Therefore, the asymptotic behaviour of costs for a process having several transient states remains the same as for a process without such states. That gives us the possibility of using Howard's algorithm also for the more general situation.

If a random process governed by some strategy has several closed subsets (which is a rare situation in practical applications), then a modification of the above algorithm should be used (see Howard [46, Chapter 6]).

4. Problems and Exercises

(1) There is a sequence of points $A_0, A_1, \ldots,$ randomly disposed on the time axis in such a way that the distances between every two adjacent points are independent random variables with a given DF $F(t)$ (see Figure 9). Denote by $N(t)$ the ME of the number of points disposed within the interval $(0, t)$ – (the point A_0 is not counted). Derive the following recurrence relation for $N(t)$:

$$N(T) = F(T) + \int_0^T N(T-t)\,\mathrm{d}F(t).$$

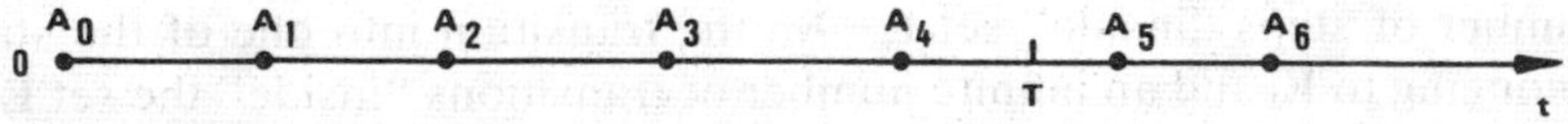

Figure 9.

(2) A Markov chain with the set of states $\mathbf{E} = \{E_1, E_2, \ldots, E_n\}$ is given. The transitions took place at regular time instants $t = 0, 1, 2, \ldots$. A payoff c_{ij} is counted for the transition from E_i into E_j. Write the expression for the expected payoff rate.

(3) For Example 2, Section 3.1, write the formula for the expected penalty per unit time for a stationary strategy which prescribes the preventive repair after the period T of failure-free operation. Set c_a and c_p as the ER and PR costs.
Answer:

$$g(\sigma) = \frac{(1-F(T))c_p + F(T)c_a}{\int_0^T [1-F(x)]\,\mathrm{d}x + (1-F(T))t_p + F(T)t_a},$$

$$t_p = \int_0^\infty t\,\mathrm{d}F_p(t), \qquad t_a = \int_0^\infty t\,\mathrm{d}F_a(t).$$

It is worthwhile to give an interpretation of this formula. Consider one cycle consisting of failure-free operation and the downtime. Then we have in the denominator the ME of the length of this cycle: the first term is the ME of the operation period, the second one is the ME of the downtime. The numerator is equal to the average penalty in a cycle.

Continuation of Number 3. Consider the age-replacement under the following modification of payoffs: for the failure-free operation period of length x a cost $a + bx^c$ is associated. (These payoffs may reflect, for example, the returns in the process of operation. Such payoffs were proposed in [76].)

Write the expression for the expected payoff per unit time.

(4) *The optimal positioning of PR points in the finite time interval.* There is an element with DF $F(t)$ of failure-free operation. At $t_0 = 0$ the element is put into operation. The element is intended to work up to the time $t = \theta$. At the prescribed times $t_1, t_2, \ldots, t_k$ $(t_0 < t_1 < \cdots < t_k < \theta)$. PRs are carried out which instantaneously renew the element. The process is stopped either when the first failure occurs or at the instant $t = \theta$. Let $V(\theta; t_1, t_2, \ldots, t_k)$ denote the ME of the total failure-free operation time and by $V(\theta; k)$ its maximal value when the PR points are located optimally. Find the value $V(\theta; k)$ and the optimal instants for PRs $t_1^*, t_2^*, \ldots, t_k^*$.

Solution. Show that $V(\theta; k)$ satisfies the following recurrence relation:

$$V(\theta; k+1) = \max_{0<t_1\leq\theta} \left[\int_0^{t_1} [1 - F(u)]\, du + [1 - F(t_1)] V(\theta - t_1; k) \right],$$

$$V(\theta; 0) = 0, \quad k = 0, 1, 2, \ldots.$$

For $k = 0$ we obtain

$$V(\theta; 1) = \int_0^{\theta} [1 - F(u)]\, du.$$

Proceeding in the same manner, we obtain

$$V(\theta; 2) = \max_{0<t_1\leq\theta_1} \left[\int_0^{t_1} [1 - F(u)]\, du + (1 - F(t_1)) \int_0^{\theta - t_1} (1 - F(u))\, du \right],$$

and so on. These relations can be used for obtaining an approximate solution.

(5) *Optimal rule for making PR when the number of PRs is restricted.* Consider an element with lifetime DF $F(t)$ switched into operation at $t = 0$. There are m elements in reserve which may be used for the PRs.

The element stops working at the moment τ when the first failure occurs. Find a rule for the optimal positioning of PR times which maximizes the expected value $M\{\tau\}$ up to the time of the first failure.

Solution. Let $V(k)$ be the maximal value of $M\{\tau\}$ when there are k elements in the reserve and the repair times are selected optimally. The following recurrence relations are valid for the values $V(k)$:

$$V(k) = \max_t \left[\int_0^t x \, dF(x) + (1 - F(t))(t + V(k-1)) \right],$$

$$k = 1, 2, \ldots,$$

$$V(0) = \int_0^\infty x \, dF(x).$$

From here it is possible to find $V(1), V(2), \ldots, V(k)$. The optimal instants $t_1^*, \ldots, t_k^*$ are those values of t which maximize the right-hand sides of these recurrence relations.

(6) *Simplification of counting costs.* Assume that the cost c_j is paid for entering state E_j. Show that this is equivalent (with respect to the cost rate) to the situation when the expected one-step cost is $\omega_j = c_j$.

Chapter 2

PREVENTIVE MAINTENANCE BASED ON DATA OF THE LIFETIME DISTRIBUTION FUNCTION

1. Operation of a Single Element for a Fixed Period of Time

1.1. General Description

The next two sections will deal with the preventive maintenance of a single-part system called "element", for which only two states may be distinguished: the failure state and the absence of failure. The DF $F(t)$ of failure-free operation time is assumed to be known. Examples A and B given in Section 1 of Chapter 1 dealt with the same assumptions.

The preventive maintenance rule depends essentially not only on the properties of DF $F(t)$ but also on the special features of the failure-signalling and detection system.

Let us discuss in more detail the role which the failure-signalling system plays. In Example A above, we considered a situation when the signal denoting failure was sent immediately after the system went into a failure state. This is more or less typical of equipment such as automatic production lines where special control and measuring devices constantly follow the state of the working tools, machine adjustment, etc.

Example B (see Section 1 of Chapter 1) considered an opposite situation when no information about failures at all is received by the repair team. This is typical for equipment which is stored and whose state can be revealed only by doing special inspections. Usually we have a situation "between" these two extremes – sometimes the information about failure is received and sometimes not. Computers serve as a good example.

Malfunctions which occur during computer operation do not manifest themselves. Often the results of computations seem plausible, although there is a hidden defect in the computer, which will be discovered only after the completion of all the computations, by performing the scheduled inspection and testing or scrutinizing the output data.

It can be asserted that the more general situation permits the following formal description. The failure occurs at some random instant τ. After a random "latent" period τ_0, the failure detecting device sends a signal about the failure with the probability α, $0 \leq \alpha \leq 1$. The value α is the characteristic of the failure-signalling system. Thus, in Example A (Section 1, Chapter 1) $\alpha = 1$, and in Example B $\alpha = 0$. In this section, we shall consider a model under the assumption that α is an arbitrary value, $0 \leq \alpha \leq 1$, and the "latent" period is zero. In Section 6, Problem 4, we shall discuss the case with a nonzero latent period whose length has a known DF.

It should be noted that the system of failure-signalling can also send a signal of "false alarm" (see the discussion of Example D). Such a possibility could be taken into account and that slightly complicates the formal treatment. One model of this sort will be treated in Problem 6 of Section 6 of this chapter.

The system's performance characteristics depend essentially on the quality of the failure detection and locating device. It may happen that a particular system has given no signal at all of failure, but during a short inspection, one is always detected and eliminated. Such a system probably will work well. On the other hand, the opposite may occur: in most cases the repair group receives a signal without any delay when the failure occurs, but in a few cases, the failure has no manifestations and an inspection cannot discover it. Thus, even after testing the computer, the failure can remain undiscovered and the repair group receives the wrong impression that the computer is in order. Experienced repairmen say that "the computer gets used to testing".

Nonreliable failure detection through inspection and testing may be interpreted in the following manner: there is a probability β that the existing failure will be revealed by carrying out the inspection. In the general case, $0 \leq \beta \leq 1$.

In the present section, we shall deal with the case of nonreliable signalling devices ($0 < \alpha < 1$) and reliable failure-locating devices ($\beta = 1$). In the next section, we shall consider the long-term operating element, discussing the general model with both nonzero α and β.

1.2. Statement of the Problem

An element with known failure-free operation time $F(t)$ is given, which has to operate a prescribed time interval $(0, \theta)$. In case of failure, a signal is sent immediately with the probability α; $1-\alpha$ is the probability that no signal will be sent at all. After the failure signal, the ER is carried out. Its duration is t_a and it renews the element. After the ER is performed, the element starts to operate again.

If during the period of time T after the last switching into operation, there was no signal of failure, then an inspection is carried out. Its duration is t_s and if there is a failure, it will be discovered with probability 1. If the failure is discovered, the additional time t_a will be spent to carry out the ER. If during the inspection, no failure is discovered, PR is carried out. Its duration is t_p and similarly to ER, it renews the element.

The repairman can select value T. This is the only "control" in the present situation. Our goal is to find an optimal rule for appointing inspections which will provide the maximal expected total failure-free operation time during the period $(0, \theta)$.

Let us formulate more precisely what is meant by "rule for appointing an inspection".

Let a regular switching-on in the operation be made at instant t_1. Appoint the next inspection at time

$$t_2 = t_1 + \Psi(t_1),$$

where $\Psi(u)$ is some nonnegative function defined for $0 \leqslant u \leqslant \theta$ and taking values in the interval $(0, \theta)$. If $\Psi(u) \leqslant \theta - u$, then such a function completely defines the service rule. In our case, therefore, $\Psi(u)$ is the "control strategy". This strategy may be identified as a second-type one (see Section 3.2, Chapter 1), because the choice of the inspection period depends on the time when the decision is made. In order to avoid some formal difficulties, we assume that at time θ an inspection is always appointed. In order to simplify the computational procedure for finding the optimal strategy, it will be assumed that all time constants in our problem are multiples of some unit interval Δ: $\theta = N\Delta$, $t_a = m_a\Delta$, $t_p = m_p\Delta$, $t_s = m_s\Delta$; N, m_a, m_s, m_p are integers.

Now we proceed with the construction of the formal model. Let us define the following states: E_1 – the element is operating; E_2 – the element is inspected, there is no failure; E_3 – the element is inspected, at the beginning of the inspection there was a nondiscovered failure; E_4 – the element is repaired.

Figure 10 shows the graph corresponding to this description. Vertices represent states, arcs indicate the possible transitions.

Let us calculate the probabilities of transitions from E_i into E_j. Assume that at instant t the element went to state E_1, and inspection period T was

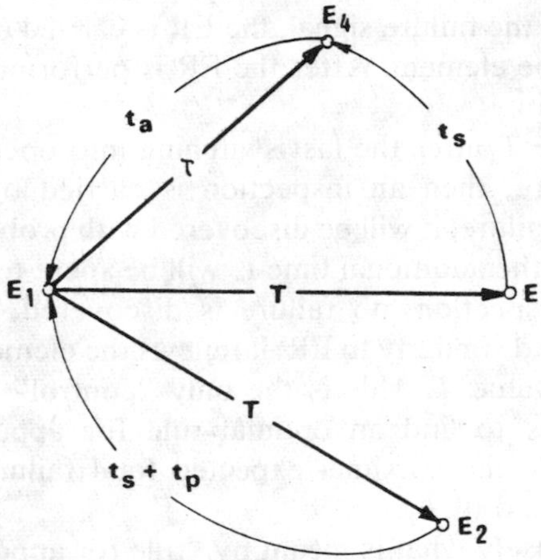

Figure 10. Transition diagram for the case $0<\alpha<1$ and $\beta=1$.

appointed. The transition $E_1 \to E_2$ will occur if and only if there is no failure during the interval $(t, t+T)$, i.e. with the probability $1-F(T)$. The transition $E_1 \to E_3$ occurs if and only if at some instant $t+\tau$, $0<\tau<T$, the failure occurs but no signal about it is received. The corresponding probability is

$$\boldsymbol{P}\{E_1 \to E_3\} = p_{13}(T) = F(T)(1-\alpha).$$

It is clear that

$$\boldsymbol{P}\{E_1 \to E_4\} = \alpha F(T) = p_{14}(T)$$

and

$$\boldsymbol{P}\{E_2 \to E_1\} = \boldsymbol{P}\{E_3 \to E_4\} = \boldsymbol{P}\{E_4 \to E_1\} = 1.$$

The matrix of transition probabilities P has the form:

$$\mathbf{P} = \mathbf{P}(T) = \left\| \begin{matrix} 0 & 1-F(T) & (1-\alpha)F(T) & \alpha F(T) \\ 1 & 0 & 0 & 0 \\ 0 & 0 & 0 & 1 \\ 1 & 0 & 0 & 0 \end{matrix} \right\|. \tag{1.1}$$

Let us point out that the future of the process under the fixed control rule does not depend on the past after a transition has been made. This is provided by the properties of repairs: they completely renew the element. Therefore, the stochastic behaviour of the random process ξ_t, with four states E_1, E_2, E_4, follows the pattern of a SMP.

Let the subsequent inspection follow the switching into operation after the interval T. Then it is easy to obtain the DFs $F_{ij}(t)$ for the SMP. Indeed,

$$F_{14}(t)=\begin{cases}\dfrac{F(t)}{F(T)}, & t<T,\\ 1, & t\geqslant T.\end{cases} \tag{1.2}$$

All other DFs are of a degenerate type:

$$\begin{aligned}&F_{13}(t)=F_{12}(t)=\delta(T),\\ &F_{34}(t)=\delta(t_s);\ F_{41}(t)=\delta(t_a);\\ &F_{21}(t)=\delta(t_s+t_p),\end{aligned} \tag{1.3}$$

where

$$\delta(y)=\begin{cases}0, & t<y,\\ 1, & t\geqslant y.\end{cases}$$

It remains to determine the returns. Let us denote the returns in such a way that the total expected return accumulated on the time interval $(0, \theta)$ would be equal to ME of the total failure-free operation time. From the definition of the process, it follows that the element is in a nonfailure state during the entire transition time $E_1 \to E_2$ and $E_1 \to E_4$ and a certain part of the transition time $E_1 \to E_3$. Therefore, the expected return associated with the transition $E_i \to E_j$ is equal to the average failure-free operation time during this transition. $F(u+\Delta u)-F(u)$ is the probability that the failure-free operation time is between u and $u+\Delta u$; $1-F(T)$ is the probability that this time is equal to T. Therefore, the expected one-step return is

$$\omega_1(T)=\int_0^T u\ \mathrm{d}F(u)+T(1-F(T)). \tag{1.4}$$

According to the definition, the one-step returns for states E_2, E_3 and E_4 have to be set zero.

1.3. Recurrence Relations for Returns

In order to simplify the notations, it will be assumed that the DF $F(t)$ is a stepwise one according to which failures can occur only at instants $k\Delta$, $k \geqslant 1$. Δ is assumed to be equal to a unit of time. Denote

$$\boldsymbol{P}\{\tau = l\Delta\} = F(l\Delta) \equiv F(l), \qquad F(0) = 0, \quad l = 0, 1, 2, \ldots;$$
$$f_l = \boldsymbol{P}\{\tau = l\Delta\} = F(l) - F(l-1), \qquad l \geqslant 1, \quad f_0 = 0. \tag{1.5}$$

Assume that if at time $t = N_1\Delta = N_1$ the failure occurs and the signal was received, ER will be carried out even though this moment was planned for carrying out an inspection. Assume that the inspection period T can be only a multiple of Δ: $T = K\Delta = K$. Hence through (1.5) the formulas for $\omega_1(T) = \omega_1(k)$, $k = 1, 2, \ldots$, and for $p_{1j}(T) = p_{1j}(k)$ take the form:

$$\omega_1(k) = \sum_{s=1}^{k} sf_s + (1 - F(k))k,$$
$$p_{12}(k) = 1 - F(k), \tag{1.6}$$
$$p_{13}(k) = (1 - \alpha)F(k),$$
$$p_{14}(k) = \alpha F(k).$$

Under the assumption mentioned, the rule for appointing inspections is completely described by a certain stepwise function $\Psi(r)$, where r is the current time and $\Psi(r) + r$ is the appointed instant of inspection. For a finite interval of time there is only a finite number of different functions Ψ and therefore among them we can always find the best one.

Let $v_1(\theta - m; \Psi)$ denote the expected total return accumulated during the interval $(\theta - m, \theta)$ if at the moment $t_1 = \theta - m\Delta = N - m$ the element was put into operation and the inspection rule was given by the function Ψ. Let $v_1(m)$ be the maximal expected return:

$$\max_{\Psi} v_1(\theta - m; \Psi) = v_1(\theta - m; \Psi^*) = v_1(m).$$

Assume that we know the values $v_1(j)$ for $j = 1, 2, \ldots, m$. Let us express $v_1(m+1)$ through them. Let a transition be made at $\tau_0 = \theta - (m+1)$ into state E_1 and the next inspection is planned to be carried out at time $\tau_1 = \theta - (m+1) + k$. $p_{1i}(k)$ is the probability that the next state will be E_i, $i = 2, 3$. In the case of $i = 2$ after time $m_p + m_s$, i.e. at the instant $\tau_2 = \theta - (m+1) + k + m_p + m_s$, the process will be again in state E_1; the expected return $v_1(m + 1 - k - m_p - m_s)$ will be accumulated if we follow the optimal inspection rule. In a similar manner, we can check that in the case of transition into E_3 the future expected return under the optimal strategy will be equal to $v_1(m+1-m_s-m_a)$.

$$p_{14}(k)\frac{F(j)-F(j-1)}{F(k)}=\alpha f_j, \qquad j \geq 1,$$

is the probability that state E_1 is followed by state E_4 after k units of time. If this happens, on the remaining interval of operation $(\theta-(m+1)+j, \theta)$, the return $v_1(m+1-j-m_a)$ will be accumulated, since at $\tau = \theta-(m+1)+j+m_a$ the ER will be completed and the system will enter state E_1. In addition, the one-step expected return for state E_1 is equal to $\omega_1(k)$. Therefore, the total expected return is equal to the value:

$$\omega_1(k)+p_{12}(k)v_1(m+1-k-m_s-m_p)$$
$$+p_{13}(k)v_1(m+1-k-m_s-m_a)+\alpha\sum_{j=1}^{k} v_1(m+1-j-m_a)f_j.$$

Now let us choose the value k so as to maximize this expression. According to the optimality principle (see Section 3.4 of Chapter 1) we obtain the desired recurrence relation:

$$\begin{aligned} v_1(m+1) = \max_{1\leq k\leq m+1} [&\omega_1(k)+p_{12}(k)v_1(m+1-k-m_s-m_p) \\ &+p_{13}(k)v_1(m+1-k-m_s-m_a) \\ &+\alpha\sum_{j=1}^{k} v_1(m+1-j-m_a)f_j]. \end{aligned} \tag{1.7}$$

Note that $v_1(l)=0$ must be assumed for $l \leq 0$. If a certain value k_0 maximizes the right side of (1.7), then the optimal CA at $\tau=\theta-(m+1)$ is $\Psi(\tau)=k_0$. The relation (1.7) is convenient for obtaining the optimal inspection rule. We demonstrate this by an example.

Example. Assume that $\theta = 10$ units of time; $m_s = m_p = 1$, $m_a = 2$; the probability of signalling the failure is $\alpha = 0.5$. All remaining data are presented in Table 1. At $t=0$ a new element starts the operation. We have to find the optimal inspection rule.

Substituting the values of m_s, m_p and m_a into (1.7) we obtain

$$\begin{aligned} v_1(m+1) = \max_{1\leq k\leq m+1} [&\omega_1(k)+p_{13}(k)v_1(m-2-k) \\ &+p_{12}(k)v_1(m-1-k)+\alpha\sum_{j=1}^{k} v_1(m-1-j)f_j]. \end{aligned}$$

Table 1

k	$F(k)$	$f_k = F(k) - F(k-1)$	αf_k	$p_{12}(k) = 1 - F(k)$	$p_{13}(k) = (1-\alpha)F(k)$	$\omega_1(k) = \Sigma_1^k sf_s + [1-F(k)]k$
1	0.070	0.070	0.035	0.930	0.035	1.000
2	0.220	0.150	0.075	0.780	0.110	1.930
3	0.410	0.190	0.095	0.590	0.205	2.170
4	0.650	0.240	0.120	0.350	0.325	3.300
5	0.830	0.180	0.090	0.170	0.415	3.650
6	0.950	0.120	0.060	0.050	0.475	3.820
7	1.000	0.050	0.025	0.000	0.500	3.870
8	1.000	0.000	0.000	0.000	0.500	3.870
9	1.000	0.000	0.000	0.000	0.500	3.870
10	1.000	0.000	0.000	0.000	0.500	3.870

Let us calculate $v_1(m)$ for $m = 1, 2, \ldots,$ by moving "back" from the end of the time interval. Set $m = 0$. It is clear that

$$v_1(1) = \max_k \omega_1(k) = \omega_1(1) = 1.$$

If the elements start to work at $t = 9$, the only possibility is $\Psi(9) = 1$. Then for $m = 1$

$$v_1(2) = \max_{1 \le k \le 2} \omega_1(k) = \omega_1(2) = 1.93; \qquad \Psi(8) = 2.$$

For $m = 2$

$$v_1(3) = \max_{1 \le k \le 3} \omega_1(k) = \omega_1(3) = 2.71; \qquad \Psi(7) = 3.$$

For $m = 3$ the expression for v_1 becomes slightly different:

$$v_1(4) = \max_{1 \le k \le 4} \Big[\omega_1(k) + p_{12}(k) v_1(2-k) + p_{13}(k) v_1(1-k) + 0.5 \sum_{j=1}^{k} v_1(2-j) f_j \Big]$$

$$= \max \begin{cases} \omega_1(1) + p_{12}(1) v_1(1) + v_1(1) \cdot 0.5 f_1 = 1.965. \\ \omega_1(2) + \qquad\qquad v_1(1) \cdot 0.5 f_1 = 1.965. \\ \omega_1(3) + \qquad\qquad v_1(1) \cdot 0.5 f_1 = 2.745. \\ \omega_1(4) + \qquad\qquad v_1(1) \cdot 0.5 f_1 = 3.335. \leftarrow \end{cases}$$

$v_1(4) = 3.335, \qquad \Psi(6) = 4.$

Similarly for $m = 4$

$$v_1(5) = \max_{1 \le k \le 5} \left[\omega_1(k) + p_{12}(k) v_1(3-k) + p_{13}(k) v_1(2-k) + \alpha \sum_{j=1}^{k} v_1(3-j) f_j \right] = 3.793,$$

$\Psi(5) = 5.$

Further calculations are carried out following the same scheme. We shall give only the results – the maximal values of v_1 and the optimal inspection periods:

$v_1(6) = 4.155; \qquad \Psi_1(4) = 6,$
$v_1(7) = 4.576; \qquad \Psi_1(3) = 2,$
$v_1(8) = 5.345; \qquad \Psi_1(2) = 3,$
$v_1(9) = 5.980; \qquad \Psi_1(1) = 3,$
$v_1(10) = 6.580; \qquad \Psi_1(0) = 4.$

Figure 11 shows both the values $v_1(m)$ and the optimal inspection period. After starting the operation, the inspection is planned at time $t = 4$. If, for example, a signal about failure was received at $t_1 = 3$, at time moment $t_1 + t_a = 3 + 2 = 5$, the element starts the operation again and the coming inspection is planned to be carried out at $t_2 = 5 + \Psi(5) = 10$, i.e. up to the end of the planned operation, no inspections should be carried out. Following the optimal strategy, we obtain $v_1(10) = 6.58$ which means approximately an efficiency of 0.66. In order to compare this result with the one which corresponds to the initially nonoptimal choice of inspection period, we present the following data:

$\tilde{\Psi}(0) = 1; \qquad \tilde{v}_1(10) = 5.56,$
$\tilde{\Psi}(0) = 2; \qquad \tilde{v}_1(10) = 6.06,$
$\tilde{\Psi}(0) = 3; \qquad \tilde{v}(10) = 6.46,$
$\tilde{\Psi}(0) = 4; \qquad \tilde{v}_1(10) = 6.58, \leftarrow$
$\tilde{\Psi}(0) = 5; \qquad \tilde{v}_1(10) = 6.39,$
$\tilde{\Psi}(0) = 6; \qquad \tilde{v}_1(10) = 5.98.$

It follows that the initial deviation from the optimal inspection period changes the efficiency by $3 \div 15\%$.

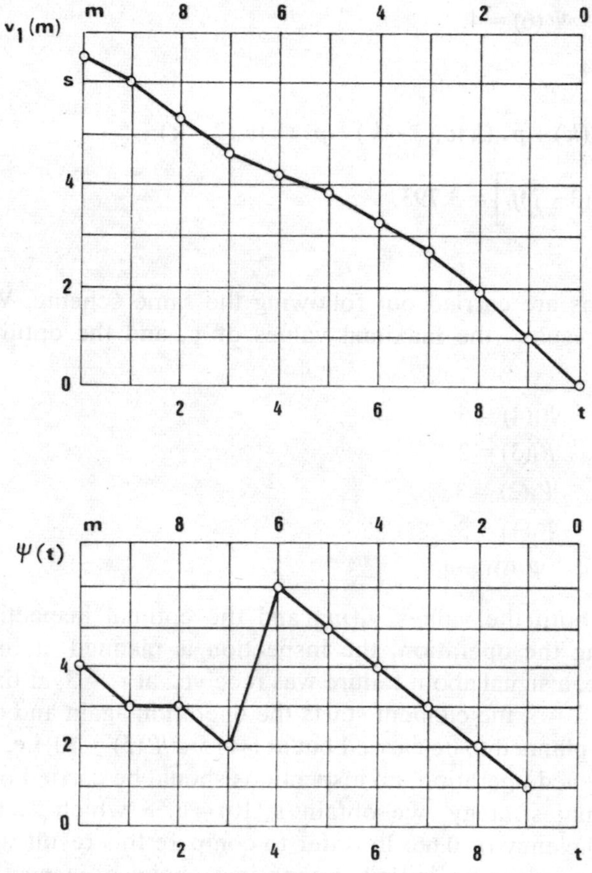

Figure 11. Optimal strategy $\Psi(t)$ and the corresponding maximal expected return $v_1(m)$. t is the current time and $m = \theta - t$ is the time up to the end of operation.

1.4. Conclusion

The problem stated in this section is a typical one for reliability theory. In the literature devoted to preventive maintenance, often a similar problem is considered with the only difference that the proposed criterion reflects "economic" performance characteristics (e.g. service cost or the gain received from the system's operation). Many problems of that sort are considered in [49, Chapters 4 and 5]. We have to note only that from the formal point of view, all these problems are very similar to the one we have just discussed in this section. The changes in the statement of the problem lead only to certain modifications of one-step returns and transition probabilities. Later on, a standard recurrence technique can be applied for obtaining the optimal strategy.

2. *Operation of a Single Element for an Infinite Period of Time*

2.1. *Statement of the Problem*

In this section we continue to discuss the model of the previous section with some modifications. The first one is that now an infinite period of operation will be considered; the second one is a nonreliable failure detection system when inspection is carried out. It will be supposed that if at the beginning of the inspection the element is in a failure state, then the inspection discovers the failure with probability β, and $0 < \beta < 1$. If during the inspection, the repair team does not discover the failure, it acts as if the element has not failed, carries out the PR of the element and puts it into operation. It is natural to suppose that if, in fact, the failure remained undiscovered, then the element is functioning although being in the failure state. In this case, no signal about the failure will be received and while carrying out the next inspection, the failure detection procedure will be repeated. In such a manner, the element will operate until an inspection finally discovers the existing failure. After this the ER will be carried out and the element will start its operation again as if entirely new.

Let us add to states $E_1 \div E_4$ defined in the previous section a new state E_5 defined as follows: the element is functioning in the presence of a nondiscovered failure. A diagram in Figure 12 shows the interaction of states $E_1 \div E_5$.

We have previously pointed out in Section 3 of Chapter 1 that for processes having an infinite length, the efficiency of preventive maintenance is usually measured by means of the expected cost or return per unit time.

If the return is defined in a manner similar to that in Section 1, the return rate will express the operational readiness.

It should be specified that in the case of an operation on an infinite time interval, we restrict ourselves to only a consideration of Markovian stationary strategies. The application of these strategies to the model under consideration is as follows: whenever the element has started to

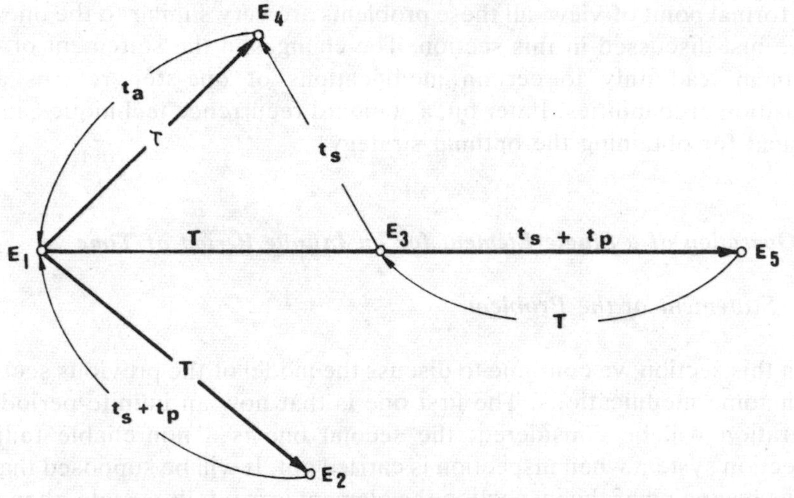

Figure 12. Transition diagram for the case of nonreliable failure signalling ($0 < \alpha < 1$) and "incomplete" inspection ($0 < \beta < 1$).

operate, the next inspection is planned after a constant period of time T. Therefore, T is the only parameter determining the strategy.

The matrix of transition probabilities has a form resembling that of (1.1). The changes are caused by the presence of a new state E_5. The only possible transitions from E_5 is into E_3. Besides that, possibilities occur to go from E_3 into E_4 with probability β (when the failure is discovered during inspection) and from E_3 into E_5 in the opposite case. The matrix **P** has the form:

$$\mathbf{P} = \mathbf{P}(T) = \left\| \begin{matrix} 0 & 1-F(T) & (1-\alpha)F(T) & \alpha F(T) & 0 & E_1 \\ 1 & 0 & 0 & 0 & 0 & E_2 \\ 0 & 0 & 0 & \beta & 1-\beta & E_3 \\ 1 & 0 & 0 & 0 & 0 & E_4 \\ 0 & 0 & 1 & 0 & 0 & E_5 \end{matrix} \right\| . \qquad (2.1)$$

2.2. Formula for Operational Readiness

In order to obtain an expression for the OR, we use the general formula (3.44) from Chapter 1. If we write T instead of σ_i, we obtain

$$g(T)=\sum_{i=1}^{5}\pi_i(T)\omega_i(T)\Big/\sum_{i=1}^{5}\pi_i(T)\nu_i(T), \tag{2.2}$$

$\pi_i(T)$ must be found from the system of equations

$$(\pi_1(T),\dots,\pi_5(T))\equiv\boldsymbol{\pi}(T)=\boldsymbol{\pi}(T)\mathbf{P}(T),$$
$$\sum_{i=1}^{5}\pi_i(T)=1. \tag{2.3}$$

The expected one-step return $\omega_1(T)$ for state E_1 was defined by (1.4):

$$\omega_1(T)=\int_0^T u\,\mathrm{d}F(u)+T(1-F(T))=\int_0^T(1-F(u))\,\mathrm{d}u.$$

The other $\omega_i(T)$ have to be set zero because in states $E_2\div E_5$ the element either has been repaired or is operating, having failed. It remains only to find $\nu_i(T)$. According to formula (3.42) from Chapter 1,

$$\nu_1(T)=\sum_{i=1}^{5}p_{1i}(T)\int_0^\infty\tau\,\mathrm{d}F_{1i}(\tau).$$

The DFs $F_{1i}(t)$ are the same as in Section 1 (see (1.2) and (1.4)). Then we obtain

$$\nu_1(T)=(1-F(T))T+(1-\alpha)F(T)T+\alpha F(T)\int_0^T\frac{u\,\mathrm{d}F(u)}{F(T)}$$
$$=\int_0^T(1-\alpha F(u))\,\mathrm{d}u. \tag{2.4}$$

The other $\nu_i(T)$ are equal:

$$\nu_2(T)=t_s+t_p;\qquad \nu_4(T)=t_a;$$
$$\nu_3(T)=\beta t_s+(1-\beta)(t_s+t_p);\qquad \nu_5(T)=T. \tag{2.5}$$

We note that the values of time spent for PR and ER appear only in the formulas for $\nu_i(T)$. They will remain the same if the times of PR and ER are random with the corresponding mean values.

Substituting the matrix (2.1) into (2.3), we obtain the system of equations

$$\begin{aligned}
\pi_1(T) &= \pi_2(T) + \pi_4(T),\\
\pi_2(T) &= (1 - F(T))\pi_1(T),\\
\pi_3(T) &= (1-\alpha)F(T)\pi_1(T) + \pi_5(T),\\
\pi_4(T) &= \alpha F(T)\pi_1(T) + \beta\pi_3(T),\\
\pi_5(T) &= (1-\beta)\pi_3(T),\\
\sum_{i=1}^{5} \pi_i(T) &= 1.
\end{aligned} \tag{2.6}$$

Its solution is the following:

$$\begin{aligned}
\pi_1(T) &= C\beta, & \pi_4(T) &= CF(T)\beta,\\
\pi_2(T) &= C\beta(1 - F(T)), & \pi_5(T) &= C(1-\alpha)(1-\beta)F(T),\\
\pi_3(T) &= C(1-\alpha)F(T), & &
\end{aligned} \tag{2.7}$$

where

$$C = [2\beta - 2\alpha F(T) + \alpha\beta F(T) + 2F(T)]^{-1}.$$

Substituting the values of ω_i, π_i and ν_i into (2.2), we obtain the desired formula:

$$\begin{aligned}
g(T) = \beta \int_0^T (1 - F(u))\,du \Big[\beta \int_0^T (1 - \alpha F(u))\,du + (1 - F(T))\beta(t_s + t_p)\\
+ F(T)((1-\alpha)(1-\beta)(t_p + T) + \beta t_a + (1-\alpha)t_s\Big]^{-1}.
\end{aligned} \tag{2.8}$$

We remind the reader of the physical meaning of value $g(T)$: it is the average rate of time in which the element is in a nonfailure state. Another interpretation of this value: $g(T)$ is the probability that at some arbitrarily taken instant of time t, $t \to \infty$, the element is operating without failure [19, Section 2.3].

The problem of finding the optimal inspection rule is reduced now to the determination of a value T^* which provides the maximal value for $g(T)$. It may happen also that $T^* = \infty$, which means that the optimal strategy is not to do inspections at all.

We suppose that the optimal T^* might be found by means of direct enumeration on the discrete time scale. For practical needs, it is enough to restrict looking for the maximum in the region $0 \leqslant T \leqslant 2.5\mu$, where μ is the ME of the failure-free operation time. The extension outside this region will not give any substantial improvement of the maximal value because at $T = 2.5$ $F(T) \simeq 1$; the expression $\int_0^T (1 - F(u))\,du$ remains constant when $T \geqslant 2.5\mu$, and the denominator in (2.8) increases rapidly with T.

2.3. Investigation of the Formula for Operational Readiness

The formula for the OR (2.8) was investigated for the following special cases of practical importance:

(A) The failure-signalling system gives notification of the failure with the probability $\alpha = 1$.
(B) There is no failure-signalling ($\alpha = 0$).
(C) The system of discovering failures during the inspection always detects the existing failure ($\beta = 1$).

The special formulas for these cases can be obtained easily from (2.8). The study of the above special cases was made for the family of gamma distributions

$$F(t) = F_k(\lambda, t) = 1 - e^{-\lambda t} \sum_{i=0}^{k-1} \frac{(\lambda t)^i}{i!}. \tag{2.9}$$

This family is widely used in reliability theory (see [41, Chapter 4]). When the parameter $k = 1$, $F(t) = 1 - e^{-\lambda t}$ – the exponential distribution. The ME and variance for the DF $F_k(\lambda, t)$ are

$$\mu = \frac{k}{\lambda}, \qquad \sigma^2 = \frac{k}{\lambda^2}. \tag{2.10}$$

The ratio

$$v = \frac{\sigma}{\mu} = \frac{1}{\sqrt{k}} \tag{2.11}$$

is called the coefficient of variation and it decreases by the increase of k. When k is large, the gamma distribution is close to the normal distribution.

The density of the gamma distributions for different values of k and λ is plotted in Figure 13.

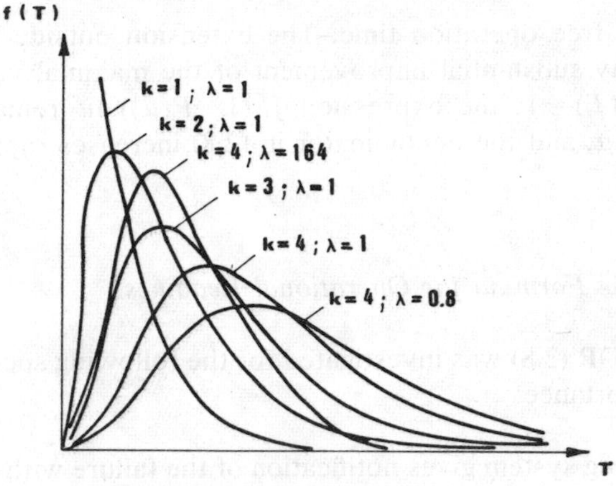

Figure 13. The density of the gamma distribution for different values of k and λ.

The gamma distribution has an increasing failure rate (by $k > 1$), see [41, Chapter 6].

We shall demonstrate that in the numerical study of formula (2.8), one parameter λ can be excluded. Indeed, according to (2.9)

$$F_k(\lambda, u) = F_k(k, u/\mu).$$

Substituting $F_k(k, u/\mu)$ instead of $F(u)$ into (2.8) and introducing a new variable $v = u/\mu$, we obtain:

$$\max_{T>0} g(T) = \max_{T_1 = T/\mu > 0} \left\{ \beta \int_0^{T_1} (1 - F_k(k, v))\, dv \right.$$
$$\times \left[\beta \int_0^{T_1} (1 - \alpha F_k(k, v))\, dv + (1 - F_k(k, T_1))(c_s + c_p)\beta \right.$$
$$\left.\left. + F_k(k, T_1)((1-\alpha)(1-\beta)(c_p + T_1) + \beta c_a + (1-\alpha)c_s) \right]^{-1} \right\}, \tag{2.12}$$

where

$$c_a = \frac{t_a}{\mu}, \qquad c_p = \frac{t_p}{\mu}, \qquad c_s = \frac{t_s}{\mu}, \qquad T_1 = \frac{T}{\mu}.$$

Therefore, it is enough to find the maximum in (2.8) setting $F(t) =$

$F_k(k, t)$, i.e. for a gamma distribution having the ME $\mu = 1$. If, in the formula (2.12), the maximum is attained at point $T_1 = T_1^*$, then the optimal inspection period is $T = T^* = T_1^* \cdot \mu$.

The maximum was sought on the discrete time lattice $T_1 =$ 0.05(0.050)2.50. Specially made calculations show that the error in the value of the maximum does not exceed 0.1–0.3%.

Special case A: ($\alpha = 1$)

When $\alpha = 1$, the preventive maintenance is carried out in the following manner.

At $t = 0$ a new element starts to operate. If up to the moment $t = T$ (T is the inspection period) the failure has not occurred, PR is started at instant $t = T$. In this special case, there is complete confidence of the absence of failure (otherwise a failure signal would have been received). Therefore, the time t_s for checking whether there is a failure or not has to be assumed equal to zero. If a failure occurs before the planned inspection at some instant t_1, $0 < t_1 < T$, ER will be carried out. After PR and ER the element is put into operation.

If the element has an increasing failure rate which means that it is subjected to aging, the effect of the above preventive maintenance is caused mainly by the fact that the time for carrying out PR is essentially smaller than the time for ER, i.e. $t_p \ll t_a$. That is why by making PRs it is possible to increase the operational readiness. Figure 14a shows the graph of states for this special case.

In the literature, the case under consideration is called the "age replacement". For this model of preventive maintenance, it is typical that the moment of carrying out the PR is not known in advance; it is made if and only if the failure-free operation period reaches some "critical" value ("age") T. Setting in (2.12) $\alpha = 1$, $c_s = 0$ we obtain

$$g(T_1)|_{\alpha=1, c_s=0} = g_A(T_1) = \int_0^{T_1} (1 - F_k(k, u))\, du \left[\int_0^{T_1} (1 - F_k(k, u))\, du + (1 - F_k(k, T_1))c_p + F_k(k, T_1)c_a \right]^{-1}. \tag{2.13}$$

Table 2 contains the results of the numerical study of this expression for different values of the parameter k and c_a. The directions for using the table are placed below it.

Let us discuss some interesting features of the age replacement.

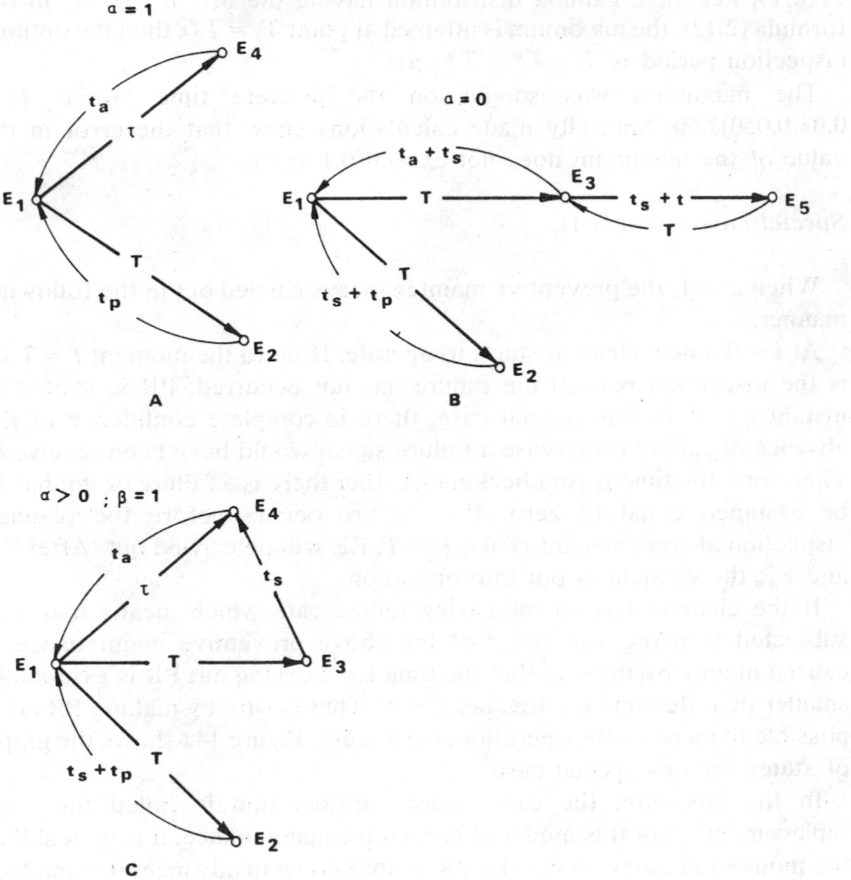

Figure 14. Transition diagram for special cases *A*, *B*, *C*.

It follows from the table that the optimal value of g increases with the increase of k. From (2.11) it follows that an increase of k means a decrease of the coefficient of variation. When $k \to \infty$ and $\mu = 1$, then the DF $F_k(k, u)$ tends to the degenerate DF having the form

$$F_\infty(\infty, u) = \begin{cases} 0, & u \leqslant 1 \\ 1, & u > 1. \end{cases}$$

For such a distribution the optimal $T_1^* = 1$, i.e. the element must be replaced immediately before it fails. For such an "ideal" DF, the OR

Table 2
Optimal parameters of preventive maintenance in the special case A: ($\alpha = 1$, $c_s = 0$).

	$c_a = 0.05$				$c_a = 0.10$				$c_a = 0.20$			
c_p	$k = 2$	$k = 4$	$k = 8$	$k = 12$	$k = 2$	$k = 4$	$k = 8$	$k = 12$	$k = 2$	$k = 4$	$k = 8$	$k = 12$
0.01	0.957 0.65	0.968 0.55	0.976 0.55	0.979 0.55	0.932 0.35	0.959 0.35	0.972 0.45	0.977 0.50	0.900 0.20	0.948 0.25	0.968 0.35	0.974 0.45
0.02	0.952 ∞	0.956 0.80	0.963 0.70	0.966 0.70	0.917 0.65	0.939 0.50	0.954 0.55	0.960 0.55	0.873 0.35	0.921 0.35	0.946 0.45	0.955 0.50
0.04					0.909 ∞	0.917 0.80	0.928 0.70	0.934 0.70	0.847 0.65	0.884 0.55	0.912 0.55	0.923 0.55
0.08									0.833 ∞	0.846 0.80	0.866 0.70	0.876 0.70

Notation and directions for use. k – parameter of the gamma distribution (2.9). $c_p = t_p/\mu$; $c_a = t_a/\mu$. The upper number in each box is the optimal value of operational readiness g_A^*. The lower number is the optimal inspection period T_1^*.

Example. Find the optimal inspection period and maximal OR for an element having $\mu = 2$, $k = 8$, $t_a = 0.2$, $t_p = 0.03$.

Solution. $c_a = 0.1$; $c_p = 0.015$; interpolating in respect to c_p between 0.01 and 0.02, we obtain

$$g_A^* = (0.972 + 0.954)/2 = 0.963; \qquad T_1^* = (0.45 + 0.55)/2 = 0.50.$$

Therefore, the optimal inspection period $T^* = 0.50 \cdot \mu = 1.00$.

will reach the value

$$g|_{F=F_\infty} = g_{\sup} = \frac{1}{1+c_p} \simeq 1 - c_p + c_p^2.$$

For the case $k = 1$ (exponential distribution), it can be shown by the direct use of formula (2.13) that the optimal $T_1 = \infty$. This means that it is not expedient at all to do preventive repairs. We want to comment on this fact which possibly is well known to the reader familiar with the topic of preventive maintenance.

An element with an exponentially distributed lifetime which has not failed up to the time instant τ_0 has the probability of failure-free operation in the interval $(\tau_0, \tau_0 + t)$, calculated under conditions of failure-free operation in the interval $(0, \tau_0)$, equal to the probability of absence of failure in the interval $(0, t)$. In that sense, the element is not subjected to aging and there is no reason to carry out its preventive repair.

As to the influence of the parameters c_a and c_p on the maximal value of g_A, we see from Table 2 that with an increase of these values, the OR decreases. This fact is intuitively clear.

Sometimes it is desirable to compare the optimal age replacement with a service policy in which no PRs are carried out. In other words, it is necessary to compare $\max_{T_1} g_A$ with $\lim_{T_1 \to \infty} g_A(T_1) = g_A(\infty)$.

There is the following relation:

$$\lim_{T \to \infty} \int_0^T (1 - F(x))\, dx = \mu,$$

where μ is the ME of the random variable with DF $F(x)$, $F(x) = 0$ (for $x < 0$). From this and from (2.13) it follows that

$$\lim_{T_1 \to \infty} g_A(T_1) = g_A(\infty) = 1/(1 + c_a).$$

Thus, for the example in Table 2 we have

$$g_A(\infty) = [1 + 0.1]^{-1} \simeq 0.91$$

and, therefore, the optimal age replacement provides a gain in the value of OR of approximately 5%.

Let us consider some general conditions under which age replacement is expedient. Let $f(t)$ denote the probability density function and $\rho(t) = f(t)/(1 - F(t))$ the failure rate. It is intuitively clear that it is not advantageous to do PR for an element which has a decreasing failure rate:

this would result in a new element whose properties are worse than the replaced one. Therefore, it is worth considering only the case of increasing failure rate. In addition, of course, a necessary prerequisite of an advantageous age replacement is that $t_p < t_a$. Assume that there is a value T such that the following condition holds:

$$\varphi(T) = \rho(T) \int_0^T (1 - F(u))\,du - \frac{t_p}{t_a - t_p} - F(T) > 0.$$

This entails an age replacement with a finite inspection period which provides a gain in comparison with case $T_1 = \infty$ (no preventive repairs). The optimal age T is a solution of the equation $\varphi(T) = 0$. The increasing failure rate ensures that this equation has only one solution.

Concluding this special age replacement case, we mention one useful property of the optimal PR "age" T^* given in [20, Chapter 4]. For every DF $F(t)$, the following inequality is valid:

$$T^* \geqslant \frac{t_p}{t_a} \mu,$$

where

$$\mu = \int_0^\infty t\, dF(t).$$

Special case B: $(\alpha = 0)$

Inspection is carried out after time T has elapsed since the element started to operate and lasts time t_s. If the inspection discovers the failure, the ER is carried out. When $\beta < 1$, there is a possibility that the inspection does not discover the existing failure. In that case, the repair team acts as if there is no failure, carries out the PR during time t_p and afterwards puts the element into operation.

The special case under consideration reflects a situation which may appear during long-term storage. An element in storage usually does not have any failure signalling (see Example B, Section 1, Chapter 1). It was mentioned above that OR may be interpreted as a probability that at some arbitrary moment t (t is large), the element is in the operable (nonfailure) state. In the case of storage, the word "operation" means "to be in storage" and "failure-free operation time" is the time when an element is in storage and in a nonfailure state.

Table 3
Parameters of the optimal preventive maintenance in the special case B: ($\alpha = 0$).

$c_p = 0$		$c_s = 0.1$						
		k						
β	c_a	1	2	3	4	6	8	12
0.8	0.1	0.576 0.35	0.683 0.40	0.731 0.40	0.760 0.45	0.792 0.50	0.809 0.50	0.830 0.55
	0.2	0.546 0.35	0.661 0.35	0.715 0.40	0.746 0.40	0.782 0.45	0.802 0.50	0.825 0.55
	0.3	0.516 0.35	0.641 0.35	0.700 0.35	0.735 0.40	0.774 0.45	0.796 0.50	0.820 0.55
	0.4	0.491 0.35	0.622 0.35	0.687 0.35	0.724 0.40	0.767 0.45	0.790 0.45	0.816 0.50
0.9	0.1	0.598 0.40	0.700 0.40	0.746 0.45	0.771 0.50	0.801 0.55	0.818 0.55	0.837 0.60
	0.2	0.564 0.40	0.676 0.40	0.726 0.40	0.756 0.45	0.790 0.50	0.808 0.55	0.829 0.60
	0.3	0.534 0.40	0.653 0.35	0.710 0.40	0.743 0.40	0.780 0.45	0.801 0.50	0.825 0.55
	0.4	0.507 0.40	0.634 0.35	0.695 0.35	0.732 0.40	0.772 0.45	0.795 0.50	0.820 0.55
1.0	0.1	0.619 0.40	0.717 0.45	0.759 0.50	0.873 0.50	0.811 0.55	0.826 0.60	0.844 0.65
	0.2	0.583 0.40	0.689 0.45	0.738 0.45	0.766 0.50	0.797 0.50	0.815 0.55	0.835 0.60
	0.3	0.550 0.40	0.666 0.40	0.719 0.40	0.751 0.45	0.786 0.50	0.806 0.50	0.828 0.55
	0.4	0.522 0.40	0.643 0.40	0.703 0.40	0.738 0.40	0.777 0.45	0.799 0.50	0.823 0.55

Table 3 (*continued*).

$c_p = 0$		$c_s = 0.2$						
		k						
β	c_a	1	2	3	4	6	8	12
0.8	0.1	0.487 0.50	0.580 0.50	0.625 0.50	0.652 0.55	0.685 0.60	0.705 0.60	0.728 0.65
	0.2	0.464 0.50	0.561 0.50	0.609 0.50	0.639 0.50	0.675 0.55	0.695 0.60	0.721 0.60
	0.3	0.444 0.50	0.544 0.45	0.594 0.50	0.627 0.50	0.665 0.55	0.688 0.55	0.716 0.60
	0.4	0.425 0.50	0.528 0.45	0.582 0.45	0.615 0.50	0.657 0.50	0.681 0.55	0.710 0.60
0.9	0.1	0.511 0.55	0.601 0.55	0.643 0.55	0.669 0.60	0.699 0.60	0.717 0.65	0.738 0.70
	0.2	0.486 0.55	0.580 0.55	0.625 0.55	0.653 0.55	0.686 0.60	0.706 0.60	0.729 0.65
	0.3	0.464 0.55	0.560 0.50	0.609 0.50	0.639 0.55	0.675 0.55	0.696 0.60	0.722 0.60
	0.4	0.443 0.55	0.542 0.50	0.594 0.50	0.626 0.50	0.665 0.55	0.680 0.55	0.716 0.60
1.0	0.1	0.534 0.55	0.621 0.60	0.662 0.60	0.686 0.65	0.714 0.65	0.730 0.70	0.749 0.75
	0.2	0.507 0.55	0.597 0.55	0.641 0.60	0.667 0.60	0.698 0.65	0.716 0.65	0.738 0.70
	0.3	0.482 0.55	0.576 0.55	0.622 0.55	0.651 0.55	0.685 0.60	0.705 0.60	0.729 0.65
	0.4	0.460 0.55	0.556 0.50	0.605 0.50	0.636 0.55	0.673 0.55	0.695 0.60	0.721 0.60

Table 3 (*continued*).

$c_p = 0$		$c_s = 0.3$						
		k						
β	c_a	1	2	3	4	6	8	12
0.8	0.1	0.431 0.60	0.513 0.60	0.554 0.60	0.579 0.60	0.610 0.65	0.630 0.65	0.652 0.70
	0.2	0.413 0.60	0.497 0.55	0.539 0.55	0.566 0.60	0.600 0.60	0.620 0.65	0.646 0.65
	0.3	0.397 0.60	0.482 0.55	0.527 0.55	0.555 0.55	0.590 0.60	0.613 0.60	0.639 0.65
	0.4	0.382 0.60	0.468 0.55	0.514 0.55	0.545 0.55	0.582 0.55	0.605 0.60	0.633 0.60
0.9	0.1	0.456 0.65	0.536 0.65	0.575 0.65	0.598 0.65	0.627 0.70	0.645 0.70	0.665 0.70
	0.2	0.436 0.65	0.517 0.60	0.558 0.60	0.583 0.65	0.614 0.65	0.633 0.65	0.656 0.70
	0.3	0.418 0.65	0.501 0.60	0.543 0.60	0.580 0.60	0.602 0.60	0.623 0.65	0.648 0.65
	0.4	0.401 0.65	0.485 0.60	0.529 0.55	0.557 0.60	0.593 0.60	0.614 0.60	0.641 0.65
1.0	0.1	0.479 0.70	0.558 0.70	0.595 0.70	0.618 0.70	0.645 0.75	0.661 0.75	0.680 0.80
	0.2	0.457 0.70	0.537 0.65	0.576 0.70	0.600 0.70	0.629 0.70	0.646 0.70	0.667 0.75
	0.3	0.437 0.70	0.518 0.65	0.559 0.65	0.584 0.65	0.615 0.65	0.634 0.70	0.657 0.70
	0.4	0.419 0.70	0.501 0.65	0.543 0.60	0.570 0.60	0.603 0.65	0.624 0.65	0.648 0.65

See opposite for Notation to Table 3.

The graph of states for special case B is shown in Figure 14b. (2.12) after setting $\alpha = 0$ takes the form:

$$g(T_1)|_{\alpha=0} = g_B(T_1)$$

$$= \beta \int_0^{T_1} (1 - F_k(k, u))\, du \, [\beta T_1 + \beta(1 - F_k(k, T_1))(c_s + c_p)$$

$$+ F_k(k, T_1)((1-\beta)T_1 + c_s + c_p + \beta(c_a - c_p))]^{-1}. \qquad (2.14)$$

The results of the numerical examination of this expression are presented in Table 3. An example illustrating its use is placed in the same table.

Let us clear up the dependence between $\max_{T_1>0} g_B$ and the parameter k. Similar to special case A, $\max g_B$ increases with the increase of k. We stress an interesting feature of the case $\alpha = 0$ and $k = 1$. In contrast to special case A, there always exists a *finite* optimal inspection period T_1 which maximizes $g_B(T_1)$. Formally, it follows from the fact that both by $T_1 = 0$ and $T_1 = \infty$, $g_B(T_1) = 0$. The physical explanation is as follows. If there is a reliable failure-signalling system, there is no need to replace the nonfailed nonaging element. If the failure-signalling system is not proper, by carrying out the PR we replace with a certain probability an element which has failed. That makes the preventive action expedient. It should be noted that for the "ideal" distribution ($k \to \infty$)

$$\max_{T_1} g_B(T_1)|_{k\to\infty} \to = \frac{1}{1 + c_s + c_p}.$$

Figure 15 shows a typical curve of the function $g_B(T_1)$.

Notation. k – parameter of the gamma distribution (2.9). $c_p = t_p/\mu$; $c_s = t_s/\mu$. β – probability of discovering the failure by carrying out the inspection. Upper number in each box is the maximal value of OR g_B^*. Lower number is the optimal inspection T_1^*. In the case $c_p \neq 0$ recalculate c_s and c_a in the following way:

$$c'_s = c_s + c_p, \qquad c'_a = c_a - c_p.$$

Use the table with these values instead of c_s and c_a.

Example. Find the optimal preventive maintenance parameters for the DF with $k = 4$; $c_s = 0.15$, $c_a = 0.20$, $c_p = 0.05$, $\beta = 0.9$.

Solution. Calculate $c'_s = 0.15 + 0.05 = 0.20$; $c'_a = 0.20 - 0.05 = 0.15$. Interpolating in respect to c_a between $c_a = 0.10$ and $c_a = 0.20$, we find that

$$g_B^* = (0.669 + 0.653)/2 \simeq 0.66; \qquad T_1^* = 0.55 \div 0.60.$$

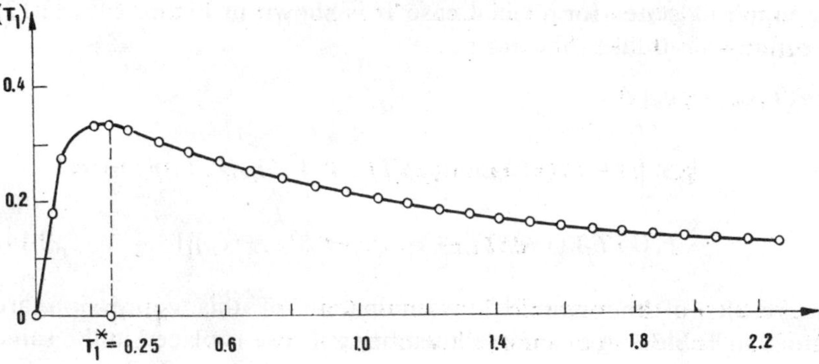

Figure 15. Dependence of the operational readiness on the inspection period for the special case B ($\alpha = 0$; $\beta = 0.3$; $c_s = 0.2$; $c_a = 0.1$; $c_p = 0$).

Special case C: $(0 < \alpha < 1, \beta = 1)$

Functioning of the element proceeds in the following way. If during time span T since the last starting of the operation no failure has occurred, an inspection is carried out. The inspection (which lasts time t_s) shows if there is a failure or not. In the first case, the inspection is followed by an ER and in the second case by a PR. After completing PR or ER the element starts to operate. This special case was considered in the previous section for the finite time interval. If $\beta = 1$, (2.8) takes the form:

$$g(T_1)|_{\beta=1} = g_C(T_1) = \int_0^{T_1} [1 - F_k(k, u)]\,du \Big[\alpha \int_0^{T_1} [1 - F_k(k, u)]\,du + (1-\alpha)T_1 + [1 - F_k(k, T_1)](c_s + c_p) + F_k(k, T_1) \times (c_a + (1-\alpha)c_s)\Big]^{-1}. \qquad (2.15)$$

The results of the numerical study of this expression are presented in Table 4.

It follows from (2.15) that $g_C(T_1) = 0$ when $T_1 = 0$ and $\lim_{T_1 \to \infty} g_C(T_1) = 0$. Therefore, similar to the case B, there is always a finite optimal inspection period T_1. As to the relation between the parameter k and the maximal value of $g_C(T_1)$, it is almost identical with case B.

From Table 4 it follows that the optimal inspection period increases

Table 4
Parameters of the optimal preventive maintenance in the special case C.

$c_s = 0.1$		$c_a = 0.1$						
		k						
α	c_p	1	2	3	4	6	8	12
0.5	0.05	0.670 0.70	0.735 0.70	0.763 0.75	0.779 0.75	0.798 0.80	0.809 0.80	0.821 0.85
	0.10	0.650 0.80	0.712 0.80	0.739 0.85	0.754 0.85	0.773 0.90	0.783 0.90	0.796 0.95
	0.15	0.633 0.85	0.643 0.90	0.720 0.90	0.735 0.95	0.754 0.95	0.764 1.00	0.778 1.00
	0.20	0.620 0.95	0.678 0.95	0.709 1.00	0.720 1.00	0.739 1.05	0.750 1.05	0.769 1.05
0.7	0.05	0.721 0.85	0.771 0.85	0.793 0.90	0.805 0.90	0.820 0.90	0.828 0.90	0.837 0.95
	0.10	0.704 0.95	0.753 1.00	0.774 1.00	0.783 1.00	0.800 1.00	0.809 1.05	0.819 1.05
	0.15	0.690 1.05	0.738 1.10	0.759 1.10	0.772 1.10	0.786 1.10	0.795 1.10	0.806 1.10
	0.20	0.678 1.15	0.726 1.15	0.747 1.15	0.760 1.15	0.775 1.20	0.785 1.20	0.797 1.20
0.9	0.05	0.803 1.35	0.832 1.25	0.844 1.25	0.851 1.25	0.859 1.20	0.863 1.20	0.869 1.20
	0.10	0.793 1.50	0.822 1.45	0.834 1.40	0.841 1.35	0.850 1.35	0.855 1.30	0.862 1.30
	0.15	0.785 1.65	0.814 1.55	0.827 1.50	0.834 1.50	0.844 1.45	0.850 1.40	0.857 1.35
	0.20	0.778 1.75	0.808 1.65	0.821 1.60	0.829 1.55	0.839 1.50	0.845 1.50	0.853 1.45

Notation. k – parameter of the gamma distribution (2.9). $c_p = t_p/\mu$; $c_a = t_a/\mu$; $c_s = t_s/\mu$. α – the probability of signalling the failure. Upper number in each box is the maximal value of OR g_C^*. Lower number is the optimal inspection period T_I^*. If $c_s \neq 0.1$,

Table 4 (*continued*).

$c_s = 0.1$		$c_a = 0.2$						
		k						
α	c_p	1	2	3	4	6	8	12
0.5	0.05	0.628 0.70	0.699 0.65	0.731 0.70	0.750 0.70	0.772 0.70	0.785 0.70	0.800 0.75
	0.10	0.610 0.80	0.676 0.75	0.705 0.80	0.723 0.80	0.743 0.80	0.755 0.80	0.769 0.80
	0.15	0.596 0.85	0.658 0.85	0.686 0.85	0.702 0.85	0.721 0.90	0.733 0.90	0.746 0.90
	0.20	0.584 0.95	0.643 0.95	0.670 0.95	0.666 0.95	0.705 0.95	0.716 0.95	0.728 1.00
0.7	0.05	0.672 0.85	0.729 0.80	0.754 0.80	0.768 0.80	0.786 0.80	0.796 0.80	0.809 0.80
	0.10	0.658 0.95	0.710 0.95	0.733 0.90	0.746 0.90	0.762 0.90	0.771 0.90	0.782 0.90
	0.15	0.646 1.05	0.696 1.05	0.717 1.00	0.730 1.00	0.745 1.00	0.754 1.00	0.764 1.00
	0.20	0.635 1.15	0.684 1.10	0.705 1.10	0.718 1.10	0.753 1.10	0.741 1.10	0.752 1.10
0.9	0.05	0.744 1.35	0.776 1.15	0.790 1.10	0.798 1.05	0.808 0.95	0.814 0.95	0.822 0.90
	0.10	0.731 1.50	0.765 1.35	0.778 1.25	0.786 1.20	0.794 1.15	0.799 1.15	0.805 1.10
	0.15	0.728 1.65	0.757 1.45	0.770 1.40	0.777 1.35	0.786 1.30	0.791 1.30	0.797 1.25
	0.20	0.722 1.75	0.751 1.60	0.764 1.50	0.771 1.45	0.780 1.40	0.785 1.40	0.791 1.35

then recalculate c_p and c_a as follows:

$$c'_p = c_p + (c_s - 0.1), \qquad c'_a = c_a + (1 - \alpha)(c_s - 0.1)$$

and use the table with these values instead of c_p and c_a.

Example. Find the optimal preventive maintenance parameters for $k = 6$ and $c_s = 0.20$,

Table 4 (*continued*).

$c_s = 0.1$		$c_a = 0.4$						
		k						
α	c_p	1	2	3	4	6	8	12
0.5	0.05	0.558 0.70	0.639 0.60	0.681 0.55	0.706 0.55	0.737 0.60	0.756 0.60	0.777 0.65
	0.10	0.544 0.80	0.616 0.70	0.652 0.65	0.674 0.65	0.701 0.65	0.718 0.65	0.738 0.70
	0.15	0.532 0.85	0.599 0.80	0.631 0.75	0.650 0.75	0.694 0.75	0.674 0.75	0.707 0.75
	0.20	0.523 0.95	0.585 0.85	0.614 0.85	0.632 0.80	0.653 0.80	0.666 0.80	0.682 0.80
0.7	0.05	0.593 0.85	0.659 0.70	0.694 0.65	0.716 0.60	0.744 0.60	0.760 0.60	0.781 0.65
	0.10	0.581 0.95	0.640 0.80	0.669 0.75	0.687 0.75	0.710 0.70	0.725 0.70	0.743 0.70
	0.15	0.572 1.05	0.625 0.90	0.651 0.85	0.667 0.85	0.686 0.80	0.698 0.80	0.714 0.80
	0.20	0.564 1.15	0.614 1.10	0.637 0.95	0.651 0.95	0.668 0.90	0.679 0.90	0.691 0.85
0.9	0.05	0.647 1.35	0.688 0.90	0.712 0.75	0.729 0.70	0.752 0.65	0.767 0.85	0.784 0.65
	0.10	0.641 1.50	0.675 1.10	0.693 0.95	0.706 0.85	0.723 0.80	0.734 0.75	0.749 0.75
	0.15	0.635 1.65	0.667 1.30	0.682 1.15	0.691 1.05	0.703 0.95	0.712 0.90	0.723 0.85
	0.20	0.631 1.75	0.660 1.35	0.674 1.30	0.681 1.20	0.691 1.10	0.697 1.05	0.705 1.00

$c_a = 0.25$, $c_p = 0.05$, $\alpha = 0.5$.

Solution. Find $c'_a = 0.30$, $c'_p = 0.15$. Interpolating in respect to c_a between $c_a = 0.40$ and $c_a = 0.20$, we find that

$$g^*_C = (0.674 + 0.721)/2 \simeq 0.70; \qquad T^*_1 \simeq 0.82.$$

with α. This has a simple explanation: the bigger α is, the greater is our confidence that there is no failure because no failure signal was given. Therefore, it is expedient to postpone the inspection.

2.4. Concluding Remarks

(1) We have previously mentioned that the choice of inspection period T was made "once and for all", without taking into consideration the "history" of the service process. It is not difficult to imagine a situation in which regarding the "past" could improve the operational readiness. Let β be a small value and α also not very large. Assume that during a long operation there was no failure, and during inspections no failure was discovered. It is hardly probable that our element really worked all the time without failures. It is more likely that there was a nonsignalled failure and the repair team simply did not discover it during inspections because value β was small. The natural conclusion is that even if an inspection does not discover the failure, it would be better to repeat it instead of setting the element into operation.

Because the failure is not always discovered during inspection, the process ξ_t with states $E_1 \div E_5$ is, in fact, only *partially* observable; the repairman cannot distinguish between states E_1 and E_5 if the failure was not detected. The situation would be desperate if in states E_1 and E_5 we had to select *different* CAs. In our case, the situation is saved because the CAs in these states are the same.

The theory of control of the partially observed process is developing very rapidly lately. We refer the reader to the work of Shiryaev [77].

By control of the partially observed process, its "history" influences the choice of the CA. In our case, apparently, this must be done by calculating the *a posteriori* probability that there is a nondiscovered failure and by acting according to its value. It would be of great interest to compare the best Markovian "memoryless" strategy with the optimal strategy based on the whole "history".

(2) We advise the reader to use Tables 2, 3 and 4 even for cases of the normal lifetime DF having parameters a (the ME) and σ^2 (the variance). This can be done by calculating the coefficient $v = a/\sigma$. Later on, one must apply the table, equating $k = 1/v^2$. Of course, there will be an error in determining the optimal period T_1^* but the error does not exceed 2–4% if $v < \frac{1}{3}$.

(3) The problem of the optimal preventive maintenance of a single-element system on an infinite time interval has often been considered in the literature. The special case A was treated in [5], [10], [13], [36] and [63] and the special case B in [64] and [83]; the general scheme for an arbitrary α, β in the case of the exponential DF was studied in [82].

3. Optimal Group Preventive Maintenance

3.1. Statement of the Problem

In previous sections we dealt only with a "system" having one element. In this section we shall deal with a system consisting of several elements of the same type which work under the same conditions. Such groups do exist in every complex system. For example, automatic production lines have groups of the same type tools, computers have many electronic devices of the same kind, etc.

For a single-element system, only two actions were available: either replace it when it fails or replace it in the absence of the failure. For a group of n elements ($n > 1$) there are possibilities to carry out both mentioned actions for every separate element and to combine preventive repairs with ER. The gain is caused by the fact that the "group" preventive repair action does not take much more time than an action carried out for an individual element. Therefore, the repair team can renew all the elements by spending a relatively small amount of time (or money) in comparison with a single-element repair. In practice, usually the ER made for the failed part of the equipment is combined with some preventive actions for other parts.

Let us proceed with the construction of the formal model. There is a group of n elements having the same lifetime DF $F(t)$. The failures of these elements are independent events. The repair team is immediately aware of the failure, i.e. the failure-signalling system always works properly.

At the initial moment $\tau = 0$, the whole group is new and has started to operate. Two constants t and T, $t < T$, are selected. The following service rule is proposed: during the period $(0, t)$ only ERs for individually failed elements are carried out; when the first failure occurs in the interval (t, T) an EPR is carried out (the ER of the failed element and the PRs for all nonfailed elements); if there was no failure in the interval (t, T), at instant T the PR for the entire group is done. EPR and PR bring the entire group to its initial state, i.e. after these actions the group may be considered as a new one. If t_1 was an instant when EPR or PR were made, then the next period $(t_1, t_1 + t)$ is appointed for ERs only, $(t_1 + t, t_1 + t + T)$ for EPR if failure occurs, and so on. Note that downtimes associated with ERs, EPR and PR are not included in the abovementioned periods $(0, t)$, (t, T), etc. (see Figure 16a). Counting time is made only over operation time without

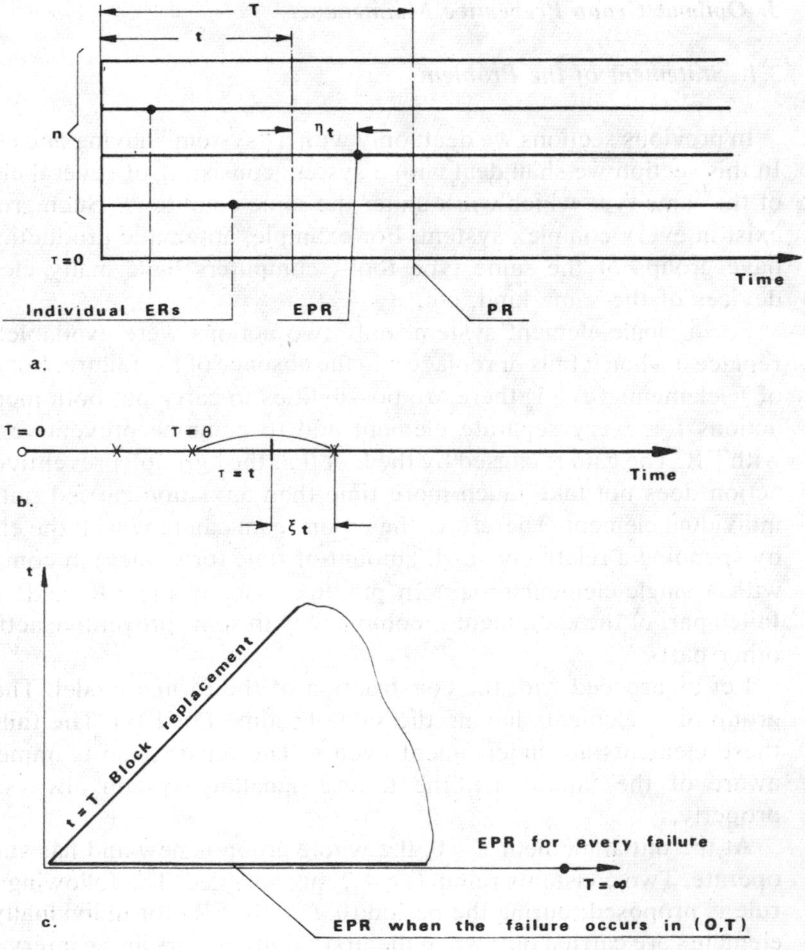

Figure 16. (a) Scheme explaining the group preventive maintenance. (b) Failure appearance in the interval (t, T). (c) "Phase" plane (T, t).

downtimes. It is assumed that downtime does not influence the lifetime of nonrepaired elements.

Every repair action is associated with some costs which we shall interpret as times needed for carrying out these actions (see the next part of this section). The problem is stated as follows. Given these costs, the DF $F(t)$, the number of elements in the group n, find out such constants $t = t_0$ and $T = T_0$ which minimize the expected average cost per unit time over an infinite operation period.

3.2. Costs Associated with ER, EPR and PR

Let t_{er}, t_{epr} and t_{pr} denote downtimes for carrying out the ER, EPR and PR respectively. Assume that there are the following simple relations between these constants:

$$\begin{aligned} t_{er} &= b_1 + b_2 = b, \\ t_{pr} &= nb\alpha, \\ t_{epr} &= b_1 + nb\alpha. \end{aligned} \tag{3.1}$$

We interpret b_1 as time spent for locating the failure; b_2 as time for replacement of one element; α is a coefficient which reflects the reduction of downtime when a group action takes place ($\alpha < 1$). As to the formula for t_{epr}, it is clear that the downtime consists of locating the failure and subsequent time for carrying out group PR.

If $n = 1$, it has to be assumed that $t_{epr} = t_{er}$ and $b\alpha = b_2$. Practically, most interesting is the case when $t_{epr} > t_{er}$, i.e. $nb\alpha > b_2$. Otherwise the problem is simplified significantly because ER always becomes less expedient than EPR. In the sequel, the values t_{er}, t_{epr} and t_{pr} are assumed to be constants; however, all will remain valid if they are random variables and the mentioned values are their MEs.

3.3. Formula for the Expected Cost per Unit Time

Let us consider one element which is immediately replaced by a new one when it fails. The DF of its lifetime is $F(t)$, $\tau_0 = 0$ is the starting point. We fix in the time axis a point $\tau = t$ and consider the random variable ξ_t which is defined as a distance from the point t to the nearest renewal point (see

Figure 16b). Let $\Psi_t(x)$ denote the corresponding DF:

$$P\{\xi_t \leqslant x\} = \Psi_t(x).$$

Consider now that we have n such elements with the same time "barrier" $\tau = t$. Let $\xi_t^{(i)}$ denote the above defined random variable for the ith element and η_t be the random variable

$$\eta_t = \min(\xi_t^{(1)}, \ldots, \xi_t^{(n)}), \tag{3.2}$$

which is the distance from the "barrier" to the nearest renewal point in the group of n elements. Let $\Psi_{nt}(x)$ be its DF. According to the well-known formula

$$P\{\eta_t \leqslant x\} = \Psi_{nt}(x) = 1 - (1 - \Psi_t(x))^n. \tag{3.3}$$

Let us introduce a time axis for our n-element system and agree upon placing on this axis only the operation periods (the downtimes are removed). Let E_1 denote the point on this axis which corresponds to EPR and E_2 the point corresponding to PR. We shall say that the system stayed in state E_i, $i = 1, 2$, the time θ, if the point E_i is followed by some point E_j occurring after the time interval θ. The words "a transition from E_1 into E_j took place" mean that the next point to E_i on the right side is E_j. The process having states E_1, E_2 is a SMP because both EPR and PR completely renew the group of n elements.

Assume that at some instant τ_0 a transition into E_i took place. Then the joint probability that the next transition will be made into state E_1 and the instant of transition will be before $\tau_0 + x + t$, is equal to

$$Q_{i1}(x+t) = \begin{cases} \Psi_{nt}(x), & \text{if } x < T - t, \\ \Psi_{nt}(T-t), & x \geqslant T - t. \end{cases} \tag{3.4}$$

Similarly, for the transition from E_i into E_2 we obtain

$$Q_{i2}(x+t) = \begin{cases} 0, & \text{if } x < T - t, \\ 1 - \Psi_{nt}(T-t), & \text{if } x \geqslant T - t. \end{cases} \tag{3.5}$$

The ME of the time between two adjacent points E_i is given by

$$\nu(t, T) = t + \int_0^{T-t} (1 - \Psi_{nt}(x))\, dx. \tag{3.6}$$

To obtain the formula for the expected cost per unit time, we note that in our SMP the transition probabilities for the states E_1 and E_2 are the same.

This follows from formulas (3.4) and (3.5). If now we take the general expression (3.44) from Chapter 1, then for our case we conclude that the denominator is equal to $\nu(t, T)$ because the mean transition times from E_1 and E_2 are the same and are equal to the mean distance between adjacent points E_i. Later on, the costs associated with the transition $E_i \to E_1$ are equal to $(nM(t)b + \alpha nb + b_1)$, where $M(t)$ is the ME of the number of renewal points for a single element in the time interval $(0, t)$. Similarly, with the transition $E_i \to E_2$ are associated costs $(nM(t)b + \alpha nb)$. From (3.4) it follows that the transition $E_i \to E_1$ occurs with the probability $\Psi_{nt}(T-t)$. The specific form of the matrix **P** (both rows are the same) provides that the stationary probabilities π_j are equal to p_{ij}: $\pi_1 = p_{i1}$, $\pi_2 = p_{i2}$. This follows immediately from system (3.45), Chapter 1. Therefore, the numerator in the formula expressing the cost rate is equal to

$$\sum_{i=1}^{2} \pi_i \omega_i = p_{i1}[nM(t)b + \alpha nb + b_1] + p_{i2}[nM(t)b + \alpha nb]$$

$$= nb\,[M(t) + \alpha + b_1 b^{-1} n^{-1} \Psi_{nt}(T-t)] = \omega(t, T). \tag{3.7}$$

Denote by $u(t, T)$ the expected cost per unit time. From (3.6) and (3.7) it follows now that

$$u(t, T) = \frac{\omega(t, T)}{\nu(t, T)} = nb \frac{M(t) + \alpha + b_1 b^{-1} n^{-1} \Psi_{nt}(T-t)}{t + \int_0^{T-t} (1 - \Psi_{nt}(x))\, dx}. \tag{3.8}$$

The values t and T are parameters which determine the preventive maintenance policy.

3.4. *Investigation of the Formula for* $u(t, T)$

The examination of (3.8) becomes simpler if the value $b_1 b^{-1} n^{-1}$ is small. This occurs if n is large ($n > 10$), $b_1 b^{-1} = 0.2 \div 0.3$. Usually $\alpha = 0.2$–0.3; $\Psi_{nt}(T-t) \ll 1$, which means that the third term in the numerator of (3.8) has the value $0.02 \div 0.01$. In that case, the increase in T influences the numerator to a small degree but does increase the denominator. Therefore, it makes sense to set $T = \infty$ and for finding the minimum of u to use the expression

$$u(t) \simeq nb\,[M(t) + \alpha + b_1 b^{-1} n^{-1}] \left[t + \int_0^{\infty} (1 - \Psi_{nt}(x))\, dx \right]^{-1}. \tag{3.9}$$

This is a simpler problem because now u depends on only one variable t. When the value $b_1 b^{-1} n^{-1}$ is small in comparison with α, the near-optimal preventive maintenance rule will be as follows: carry out ER in the time span $(0, t)$; when the first failure occurs after instant t, carry out EPR.

If we cannot do the above simplifications, we have to manage with functions $M(t)$ and $\Psi_t(x)$ which usually do not permit a simple analytic form.

$M(t)$ is called a renewal function and is equal to the following series [23, Chapter 4]:

$$M(t) = \sum_{k=1}^{\infty} F^{(k)}(t),$$

where $F^{(k)}(t)$ is the kth convolution of the DF $F(t)$, $F^{(1)}(t) = F(t)$. From here simple upper and lower estimates follow:

$$\sum_{r=1}^{k} F^{(r)}(t) < M(t) < \sum_{r=1}^{k-1} F^{(r)}(t) + \frac{F^{(k)}(t)}{1 - F(t)}. \tag{3.10}$$

Preliminary numerical computations show that the minimum in (3.8) is usually attained at such values of t which are less than $\int_0^\infty u \, \mathrm{d}F(u) = \mu$.

For the range $(0, \mu)$ a quite good estimation may be achieved if in (3.10) we take $k = 2$. (The reader is referred to a very good approximation of the renewal density function proposed by Bartholomew [12].)

The expression for $\Psi_t(x)$ might be derived by means of renewal theory:

$$\Psi_t(x) = F(t + x) - F(t) + \int_0^t h(t - z)[F(x + z) - F(z)] \, \mathrm{d}z, \tag{3.11}$$

where $h(t)$ is the renewal density function, $h(t) = \mathrm{d}M(t)/\mathrm{d}t$. If we substitute $f(t) = F'(t)$ instead of $h(t)$ and remember that $h(t) \geq f(t)$, we obtain a lower estimate for $\Psi_t(x)$.

Assume that the last renewal up to instant t took place at θ (see Figure 16c); then

$$\boldsymbol{P}\{\xi_t \leq x\} = \boldsymbol{P}\{\boldsymbol{Y} \leq t - \theta + x \,|\, \boldsymbol{Y} > t - \theta\} = \frac{F(t - \theta + x) - F(t - \theta)}{1 - F(t - \theta)}, \tag{3.12}$$

where $\boldsymbol{Y}$ is the lifetime of an element and $F(x) = \boldsymbol{P}\{\boldsymbol{Y} \leq x\}$. If it will be assumed that $F(t)$ has an increasing failure rate, then it can be proved that the ratio on the right-hand side of (3.12) increases with $(t - \theta)$, and attains its maximal value at $\theta = 0$. Therefore, we obtain the upper estimate

$$P\{\xi_t \leqslant x\} \leqslant [F(t+x) - F(t)] \cdot [1 - F(t)]^{-1}. \tag{3.13}$$

Using the inequalities (3.10) for $k = 2$, and through (3.13) we obtain the following upper estimate for $u(t, T)$:

$$u(t, T) \leqslant nb \frac{F(t) + F^{(2)}(t)[1 - F(t)]^{-1} + \alpha + b_1 b^{-1} n^{-1}\left[1 - \left(\dfrac{1 - F(T)}{1 - F(t)}\right)^n\right]}{t + (1 - F(t))^{-n} \displaystyle\int_0^{T-t} (1 - F(x+t))^n \, dx} . \tag{3.14}$$

Special calculations made show that the error resulting from the right-hand side of (3.14) does not exceed 2–3% and, therefore, this expression might be used for the numerical evaluation of optimal preventive maintenance parameters providing the minimum for $u(t, T)$.

3.5. Numerical Study of the Group Preventive Maintenance

Tables 5-1, 5-2 and 5-3 contain the results of a computerized numerical study of the approximate formula (3.14). The computation was carried out for the family of gamma distributions (see (2.9)–(2.11)).

Directions for using the tables follow:

(1) Every table is calculated for one fixed value of k (the parameter of the gamma distribution). This value is written at the head of the table.

(2) The input data for the use of the table are three parameters: α, β and n. n is the number of elements in the maintained group; $\alpha = t_{\text{pr}}/nt_{\text{er}}$; $\beta = (t_{\text{epr}} - t_{\text{pr}})/t_{\text{er}}$.

(3) The output data are three parameters: t_0, T_0 and η. t_0 is the optimal period intended for carrying out only individual ERs;

T_0 is the optimal operation age, at which the group PR must be carried out.

Both t_0 and T_0 are expressed in units of the ME of failure-free operation time of one element.

η is a coefficient which estimates the decrease in the expected cost rate using the optimal group policy in comparison with the rate obtained when no preventive actions are carried out:

$$\eta = \min_{\substack{0 \leqslant t < \infty \\ t \leqslant T}} u(t, T)/u(\infty, \infty)$$

Table 5-1.
Parameters of the optimal group preventive maintenance.
$k = 2$

		$\alpha = 0.1$			$\alpha = 0.2$			$\alpha = 0.3$	
n	β	0.3	0.5	0.7	0.3	0.5	0.7	0.3	0.5
	t_0	0	0	0	0	0	0	0	0
2	T_0	0.60	0.55	0.50	2.00	1.25	0.80	2.00	2.00
	η	0.40	0.53	0.64	0.56	0.72	0.86	0.72	0.88
	t_0	0	0	0.05	0.05	0.15	0.30	0.25	0.50
4	T_0	0.55	0.90	0.45	2.10	2.00	0.90	2.10	2.05
	η	0.44	0.55	0.66	0.68	0.80	0.90	0.91	0.99
	t_0	0	0	0.05	0.30	0.35	0.50	0.75	
8	T_0	0.55	0.50	0.45	2.20	2.10	1.00	2.00	
	η	0.52	0.61	0.71	0.83	0.88	0.93	1	
	t_0	0.10	0.15	0.15	0.40	0.50	0.60		
12	T_0	0.55	0.50	0.45	2.00	1.75	1.00		
	η	0.60	0.68	0.75	0.88	0.91	0.94		
	t_0	0.15	0.15	0.20	0.50	0.55	0.65		
16	T_0	0.60	0.55	0.40	1.40	1.45	1.00		
	η	0.67	0.74	0.81	0.90	0.92	0.94		
	t_0	0.15	0.20	0.25	0.55	0.60	0.75		
20	T_0	0.75	0.50	0.45	1.30	1.35	1.00		
	η	0.73	0.79	0.85	0.91	0.93	0.94		

Table 5-2.

$k = 6$

n	β	$\alpha = 0.1$			$\alpha = 0.2$			$\alpha = 0.3$			$\alpha = 0.4$			$\alpha = 0.5$	
		0.3	0.5	0.7	0.3	0.5	0.7	0.3	0.5	0.7	0.3	0.5	0.7	0.3	0.5
2	t_o	0	0	0	0	0	0	0	0	0	0	0	0	0	0
	T_o	0.55	0.45	0.40	0.75	0.55	0.50	0.90	0.65	0.60	1.05	0.80	0.65	1.25	0.90
	η	0.25	0.29	0.31	0.42	0.49	0.53	0.57	0.65	0.71	0.71	0.81	0.88	0.84	0.95
4	t_o	0	0	0.05	0	0	0	0.20	0.25	0.35	0.30	0.35	0.45	0.40	0.45
	T_o	0.55	0.45	0.40	0.75	0.60	0.50	0.95	0.70	0.55	0.95	0.70	0.60	1.15	0.80
	η	0.26	0.29	0.32	0.44	0.49	0.54	0.61	0.67	0.73	0.77	0.84	0.90	0.92	0.99
8	t_o	0.05	0.10	0.10	0.25	0.30	0.35	0.35	0.40	0.45	0.45	0.50	0.50	0.55	
	T_o	0.55	0.45	0.40	0.80	0.60	0.55	1.00	0.70	0.60	1.25	0.85	0.70	1.60	
	η	0.27	0.30	0.32	0.47	0.51	0.54	0.67	0.70	0.94	0.83	0.87	0.91	0.99	
12	t_o	0.10	0.20	0.25	0.30	0.35	0.30	0.45	0.45	0.45	0.50	0.55	0.55		
	T_o	0.55	0.50	0.45	0.75	0.60	0.55	1.00	0.70	0.65	1.30	0.85	0.70		
	η	0.27	0.30	0.32	0.49	0.52	0.55	0.69	0.72	0.94	0.86	0.89	0.91		
16	t_o	0.20	0.25	0.25	0.35	0.35	0.40	0.45	0.45	0.50	0.55	0.55	0.55		
	T_o	0.50	0.50	0.45	0.80	0.65	0.55	1.10	0.70	0.65	1.20	0.85	0.70		
	η	0.28	0.31	0.32	0.50	0.53	0.55	0.70	0.72	0.75	0.87	0.90	0.92		

Table 5-2 (*continued*).

		$\alpha = 0.1$			$\alpha = 0.2$			$\alpha = 0.3$			$\alpha = 0.4$		
n	β	0.3	0.5	0.7	0.3	0.5	0.7	0.3	0.5	0.7	0.3	0.5	0.7
	t_0	0.25	0.25	0.30	0.40	0.40	0.40	0.50	0.50	0.50	0.55	0.60	0.60
20	T_0	0.55	0.50	0.45	0.85	0.65	0.55	1.15	0.85	0.70	1.15	1.00	0.75
	η	0.29	0.31	0.33	0.51	0.53	0.55	0.71	0.73	0.75	0.89	0.90	0.92

Table 5-3.

$k = 12$

		$\alpha = 0.2$			$\alpha = 0.3$			$\alpha = 0.4$			$\alpha = 0.5$			$\alpha = 0.6$	
n	β	0.3	0.5	0.7	0.3	0.5	0.7	0.3	0.5	0.7	0.3	0.5	0.7	0.3	0.5
	t_0	0.15	0	0.05	0.15	0.10	0.25	0.15	0.05	0.05	0.20	0.30	0.40	0.05	0.05
2	T_0	0.70	0.60	0.55	0.80	0.65	0.65	0.85	0.75	0.65	0.95	0.80	0.70	0.95	0.80
	η	0.36	0.39	0.41	0.50	0.55	0.58	0.63	0.69	0.73	0.76	0.83	0.88	0.89	0.97
	t_0	0.15	0.05	0.30	0.25	0.05	0.40	0.35	0.40	0.45	0.40	0.45	0.50	0.45	0.50
4	T_0	0.70	0.60	0.60	0.80	0.65	0.65	0.85	0.70	0.65	0.95	0.80	0.70	1.00	0.85
	η	0.36	0.39	0.41	0.51	0.55	0.58	0.66	0.71	0.74	0.80	0.85	0.89	0.93	0.99
	t_0	0.30	0.05	0.40	0.40	0.45	0.45	0.50	0.50	0.55	0.55	0.55	0.55	0.60	

8	T_0	0.70	0.60	0.60	0.80	0.70	0.65	0.85	0.70	0.70	1.10	0.85	0.80	1.15
	η	0.37	0.40	0.42	0.53	0.57	0.59	0.69	0.72	0.75	0.84	0.87	0.90	0.98
12	t_0	0.40	0.40	0.45	0.45	0.50	0.50	0.55	0.55	0.55	0.60	0.60	0.60	
	T_0	0.70	0.65	0.60	0.80	0.70	0.65	1.00	0.80	0.70	1.10	0.85	0.80	
	η	0.38	0.40	0.42	0.55	0.57	0.59	0.71	0.73	0.75	0.86	0.88	0.90	
16	t_0	0.40	0.45	0.45	0.50	0.50	0.50	0.65	0.55	0.55	0.60	0.60	0.60	
	T_0	0.70	0.65	0.60	0.80	0.70	0.65	0.95	0.80	0.70	1.00	0.80	0.70	
	η	0.39	0.41	0.42	0.56	0.58	0.59	0.72	0.74	0.76	0.87	0.89	0.91	
20	t_0	0.45	0.45	0.45	0.50	0.50	0.55	0.55	0.55	0.60	0.60	0.60	0.65	
	T_0	0.70	0.65	0.60	0.80	0.70	0.70	0.95	0.80	0.70	1.00	0.80	0.70	
	η	0.39	0.41	0.42	0.56	0.58	0.60	0.73	0.74	0.76	0.88	0.90	0.91	

Example. An element has a gamma distribution of the failure-free operation time. The parameters are: ME $\mu = 70$ hours, the variance $\sigma^2 = 400$ hours^2. The group has $n = 12$ elements; $t_{pr} = 0.6$ hour; $t_{er} = 0.1$ hour; $t_{epr} = 0.65$ hour.

According to formula (2.11),

$$k = \mu^2/\sigma^2 = 4900/400 = 12.2.$$

Set $k = 12$. $\alpha = t_{pr}/nt_{er} = 0.6/12 \cdot 0.1 = 0.5$;

$$\beta = (t_{epr} - t_{pr})/t_{er} = (0.65\text{–}0.6)/0.1 = 0.5.$$

Applying Table 5-3, we find $t_0 = 0.60$, $T_0 = 0.85$, $\eta = 0.88$. The optimal preventive maintenance parameters expressed in hours are:

$$t_0^* = 0.60\mu = 42 \text{ hours}; \qquad T_0^* = 0.85\mu \simeq 60 \text{ hours}.$$

3.6. Special Cases of Group Preventive Maintenance

The group preventive maintenance under consideration is determined by means of two parameters t and T, where $0 \leq t \leq T$. Geometrically, every point (t, T) in the triangle area formed by the T axis and the bisector $t = T$ corresponds to a certain preventive maintenance strategy (see Figure 16c). It appears that some well-known methods of preventive maintenance can be obtained from our general group model when the area of seeking the optimal parameters is reduced to a one-dimensional area. We consider below three such special cases. In addition, it is interesting to learn what the decrease is in the efficiency of the optimal preventive maintenance in a special case, in comparison with the two-parameter general model. We shall answer this question on the basis of some numerical data.

Special case 1: $(T = t)$ *Block replacement*

Let us set $T = t$ in (3.8). Then we obtain

$$u(t, T)|_{T=t} = u_1(t) = \frac{M(t) + \alpha}{t} nb. \tag{3.15}$$

Physically, the condition $t = T$ means that we refuse to carry out EPRs. After time t since the group has started to operate, a PR is carried out. Before this moment only individual ERs have been made for failing elements. On the "phase" plane (t, T) this corresponds to the bisector $t = T$

(see Figure 16c). In the literature devoted to preventive maintenance, this method is well known and it is called "block replacement". It has been studied in several papers[5], [10], [20], [31], [36], [51] and [63]. Its popularity may be explained by its simplicity: the repair team has a clock-counter of systems operation time and starts the PR if and only if the operation time becomes a multiple of a certain prescribed constant T. We shall demonstrate that the optimal choice of the block replacement period provides the expected cost per unit time which often is near optimal to the cost in the general two-parameter model.

Table 6 presents data concerning the optimal block replacement period for the family of gamma distributions. It contains directions for finding the optimal period.

Special case 2: $(t = 0)$ *Age replacement*

Let us set $t = 0$ in formula (3.8). Recalling that $\Psi_0(x) = F(x)$ we obtain

$$u(t, T)|_{t=0} = u_2(T) = \frac{\alpha + b_1 b^{-1} n^{1}[1 - (1 - F(T))^n]}{\int_0^T (1 - F(x))^n \, dx}. \tag{3.16}$$

Physically, condition $t = 0$ means that the preventive maintenance is carried out according to the following rule. If there were no failures in the interval $(0, T)$, then the entire system is repaired at $\tau = T$. If a failure occurred in the interval $(0, T)$, then the EPR is carried out. When may such preventive servicing be profitable? Probably when the costs of individual ERs are comparable with costs of EPR. Such a situation may appear when the greatest amount of time for making an ER is spent in locating the damaged part and disassembling the unit and the least time is spent directly for the exchange of the failed element. Thus, in repairing engines, usually the individual ERs are not carried out; the repair team tries to combine the ER of the damaged part with preventive repairs for other parts.

It is easy to verify that in the case of one element in the group $(n = 1)$, this special case is reduced to the model referred to above as "age replacement" (see Section 2, special case A). The difference between formulas (3.16) and (2.13) is that the first expresses the expected losses per unit operation time, but the second, the ratio of an operation time over total time in use. Values $g_A(T)$ and $u_2(T)$ have the following relation:

$$g_A(T) = \frac{1}{1 + u_2(T)}.$$

Table 6
Parameters of the optimal block replacement.

α	k = 2	3	4	5	6	7	8	9	10	11	12	14	16	18	20
0.15	0.42 0.90	0.38 0.70	0.40 0.58	0.41 0.51	0.44 0.46	0.45 0.43	0.47 0.40	0.48 0.38	0.50 0.36	0.51 0.35	0.52 0.34	0.54 0.32	0.56 0.31	0.58 0.29	0.59 0.28
0.25		0.50 0.92	0.49 0.81	0.49 0.73	0.51 0.68	0.52 0.63	0.53 0.60	0.54 0.58	0.55 0.56	0.56 0.54	0.57 0.52	0.59 0.50	0.60 0.48	0.62 0.46	0.63 0.45
0.35			0.57 0.99	0.56 0.92	0.56 0.86	0.57 0.82	0.58 0.78	0.59 0.75	0.60 0.73	0.61 0.71	0.61 0.69	0.62 0.66	0.64 0.64	0.65 0.62	0.66 0.61
0.40					0.59 0.94	0.60 0.90	0.60 0.87	0.61 0.84	0.62 0.81	0.62 0.79	0.63 0.77	0.64 0.74	0.65 0.72	0.66 0.70	0.67 0.68
0.45						0.62 0.99	0.62 0.95	0.63 0.92	0.63 0.89	0.64 0.87	0.64 0.85	0.65 0.82	0.66 0.79	0.67 0.77	0.68 0.76
0.50								0.64 0.99	0.65 0.97	0.65 0.95	0.66 0.93	0.67 0.90	0.68 0.87	0.68 0.85	0.69 0.83
0.55												0.68 0.97	0.69 0.94	0.69 0.92	0.70 0.90
0.60														0.71 0.99	0.71 0.97

Notation. k – the parameter of gamma distribution (2.9). The upper number in each box is the optimal period of block replacement in units of the expected lifetime; the lower number is the index η = min cost rate/cost rate when no PRs are carried out.

On the "phase" plane (t, T), the axis T corresponds to the age replacement (see Figure 16c).

Special case 3: $(t = 0, T = \infty)$ *EPR is made at every failure*

If we set $t = 0$, $T = \infty$ in (3.8), then we obtain the formula

$$u(t, T)\Big|_{\substack{t=0\\T=\infty}} = u_3 = \frac{b_1 + nb\alpha}{\int_0^\infty (1 - F(x))^n \, dx} = \frac{t_{epr}}{\int_0^\infty (1 - F(x))^n \, dx}. \quad (3.17)$$

This formula corresponds to the situation when EPR is carried out after every failure. This kind of preventive service may be advantageous when values t_{epr} and t_{pr} are close. If $t_{epr} < t_{pr}$, then, of course, EPR is always more profitable than PR. Thus, when working tools are maintained in an automatic line with a special built-in block tool exchange, the time for effecting a total change of the group when one individual tool has failed, is almost the same as the time for group PR.

3.7. Additional Comments on the Model of Group Preventive Maintenance

3.7.1. The gain in the cost rate in the general two-parameter model

It is interesting to compare the optimal preventive maintenance in the above special cases with the optimal strategy in the general case.

The numerical analysis carried out for the family of gamma distributions leads to the following conclusions:

when the number of elements in the group is small ($n < 12$), then the near optimal is the age replacement ($t = 0$, special case 2); the increase in the cost rate does not exceed 3–5%.

when $n > 16$ and β is large ($\beta \geqslant 0.7$), the near optimal is the block replacement ($t = T$, special case 1); the increase in the cost rate is 2–5%.

when $n \geqslant 12$ and β is small ($\beta = 0.1 \div 0.4$), the general two-parameter case provides an essential decrease in expected costs in comparison with the best policies in special cases. The advantage of the general model using two parameters becomes more appreciable for "bad" DFs which have small values of k, and for small values of α ($\alpha \leqslant 0.2$).

Table 7 demonstrates typical relationships which exist between the values η in different preventive maintenance schemes. The smaller the index η, the more profitable is the preventive maintenance (recall the definition of this index). Table 7 illustrates also the fact that for the DFs having small variance (large k), this index is smaller.

Table 7
Comparison of different group preventive maintenance variants.

$\alpha = 0.3, \beta = 0.5$				
Parameter k of gamma distribution	$n = 4$		$n = 16$	
	min η general case	min η special cases	min η general case	min η special cases
4	0.771	0.776_2	0.842	0.907_1
8	0.617	0.618_2	0.655	0.688_2
12	0.553	0.553_2	0.577	0.591_2
16	0.516	0.517_2	0.534	0.541_2

Notations. The index to min η denotes the optimal special case: 1 – block replacement ($t = T$), 2 – age replacement ($t = 0$).

3.7.2. The random choice of preventive maintenance times

In all above models we considered only the deterministic choice of time for carrying out preventive actions. In some practical situations, the random choice of time appears in a very natural way. For example, there is a random delay in the arrival of the repairman for doing a planned repair. We shall demonstrate by one simple example how to handle this complication. Consider the example of the age replacement of a single element (see special case A, Section 2). Assume that the "critical" age T is a random variable with DF $\Phi(t)$ which does not depend on the failure-free operation time τ. Let t_p and t_a remain the times for carrying out PR and ER. According to Figure 14a, the transition $E_1 \to E_2$ occurs if it happens that $\tau > T$. Therefore,

$$p_{12} = \mathbf{P}\{\boldsymbol{\tau} > T\} = \int_0^\infty [1 - F(x)]\, \mathrm{d}\Phi(x).$$

The probability of the event "the system went from E_1 to E_2, the transition time was less than t" is given by

$$Q_{12}(t) = \mathbf{P}\{E_2, \boldsymbol{\tau}(1,2) \leq t \mid E_1\} = \int_0^t [1 - F(x)]\, \mathrm{d}\Phi(x).$$

From this follows the DF of the transition time $\boldsymbol{\tau}(1, 2)$ (calculated under the condition that this transition took place):

$$F_{12}(t) = \frac{Q_{12}(t)}{Q_{12}(\infty)} = \frac{\int_0^t [1 - F(x)]\, \mathrm{d}\Phi(x)}{\int_0^\infty [1 - F(x)]\, \mathrm{d}\Phi(x)}. \tag{3.18}$$

It is easy to verify that the ME of the length of one step starting from E_1 is equal to

$$\nu_1 = \mathbf{M}\{\min(\boldsymbol{\tau}, T)\} = \int_0^\infty [1 - F(x)][1 - \Phi(x)]\, \mathrm{d}x. \tag{3.19}$$

Using integration by parts, the last expression can be rewritten as follows:

$$\nu_1 = \int_0^\infty \left\{ \int_0^T [1 - F(x)]\, \mathrm{d}x \right\} \mathrm{d}\Phi(T). \tag{3.20}$$

Now we are ready to write the expression for the operational readiness (compare with (2.13)):

$$\tilde{g}_A(T) = \nu_1 \left[\nu_1 + t_p \int_0^\infty [1 - F(x)]\, \mathrm{d}\Phi(x) + t_a \left[1 - \int_0^\infty [1 - F(x)]\, \mathrm{d}\Phi(x) \right] \right]^{-1}. \tag{3.21}$$

The same expression can be rewritten in a more convenient form:

$$\tilde{g}_A(T) = \frac{\int_0^\infty \left[\int_0^T (1-F(x))\,dx\right] d\Phi(T)}{\int_0^\infty \left[t_p(1-F(T)) + t_a F(T) + \int_0^T (1-F(x))\,dx\right] d\Phi(T)}. \quad (3.22)$$

The question arising at this point is whether an "artificially" introduced randomization can improve the performance characteristics of preventive maintenance or not. Paper [8] contains the proof of the fact that no improvement can be obtained by the random choice of a preventive repair period. We give this proof below.

Let $\varphi(x)$ and $\Psi(x)$ be two nonnegative functions and the maximum of their ratio is attained at the point T_0:

$$\max_{0 \le x < \infty} \frac{\varphi(x)}{\Psi(x)} = \frac{\varphi(T_0)}{\Psi(T_0)}. \quad (3.23)$$

In other words,

$$\varphi(x)\Psi(T_0) \le \Psi(x)\varphi(T_0).$$

Multiplying both sides of the last inequality by $d\Phi(x)$ and integrating from 0 to ∞, we obtain

$$\Psi(T_0)\int_0^\infty \varphi(x)\,d\Phi(x) \le \varphi(T_0)\int_0^\infty \Psi(x)\,d\Phi(x),$$

or

$$\frac{\varphi(T_0)}{\Psi(T_0)} \ge \frac{\int_0^\infty \varphi(x)\,d\Phi(x)}{\int_0^\infty \Psi(x)\,d\Phi(x)}. \quad (3.24)$$

From that follows the inequality

$$\tilde{g}_A(T) \le g_A^*(T) = \max_{0 \le T < \infty} \left[\int_0^T (1-F(x))\,dx\right] \cdot \left[t_a F(T) + t_p(1-F(T)) + \int_0^T (1-F(x))\,dx\right]^{-1}, \quad (3.25)$$

which proves that the optimal randomization is provided by a degenerate function, i.e. by a constant replacement age.

3.7.3. A nonhomogeneous group

A case of practical interest is one where a group contains elements with different DFs, i.e. the ith element has a DF $F_i(t)$, $i = 1, 2, \ldots, n$. Assume that the preventive maintenance rule remains unchanged and check which changes must be done in the formula for the expected costs.

Denote by $\xi_t^{(i)}$ the time from the instant t till the first failure after that moment, for the ith element. Then the random variable

$$\tilde{\eta}_t = \min(\xi_t^{(1)}, \ldots, \xi_t^{(n)}) \tag{3.26}$$

has a DF

$$\tilde{\Psi}_{nt}(x) = 1 - \prod_{i=1}^{n}(1 - \Psi_t^{(i)}(x)), \tag{3.27}$$

where

$$\Psi_t^{(i)}(x) = \boldsymbol{P}\{\xi_t^{(i)} \leqslant x\}. \tag{3.28}$$

Formula (3.6) will be the same except for the expression of $\tilde{\Psi}_{nt}(x)$ instead of (3.3). The formula for $\Sigma_1^2\, \pi_i \omega_i$ (see (3.7)) now has the form

$$\tilde{\omega}(t, T) = nb\left[\sum_{i=1}^{n} \frac{M_i(t)}{n} + \alpha + b_1 b^{-1} n^{-1} \tilde{\Psi}_{nt}(T - t)\right], \tag{3.29}$$

where $M_i(t)$ is the renewal function for the ith element. The rest of the calculations and the methods of approximate numerical analysis remain unchanged.

It should be noted that the presence in the maintained group of 10–20% of elements which have "bad" DFs (e.g. exponential functions or DFs with decreasing failure rate) essentially decreases the efficiency of the preventive maintenance policy.

4. Which Preventive Maintenance is Better: Age or Block Replacement?

In this section we deal with the following problem. There is a single element operating over an infinite period of time. The preventive and emergency repairs are assumed to be carried out simultaneously. The costs for PR and ER are c_1 and $c_a = 1$ ($c_1 < 1$). From Sections 2 and 3 it follows that the expected cost per unit time for the age replacement is:

$$u_1(T) = [F(T) + (1 - F(T))c_1] \Big/ \int_0^T (1 - F(u))\, du, \tag{4.1}$$

where T is the "critical" age.

For a block replacement having the period of PR equal to T, the cost rate is given by the formula (see (3.15)):

$$u_2(T) = [M(T) + c_1]/T. \tag{4.2}$$

The case in question is a comparison of the following values:

$$\min_{T \geq 0} u_1(T) \quad \text{and} \quad \min_{T \geq 0} u_2(T).$$

It should be mentioned that this question is theoretical rather than practical. First, block replacement has an advantage over age replacement, because the PR times are set in advance and therefore, they can be planned beforehand. This results in the PR in block replacement usually costing *less* than the same PR in age replacement. Secondly, if we have to solve this question for a given practical situation, we can use a computer to carry out the numerical comparison to provide us with the necessary answer. We present this comparison only with the aim of illustrating how efficient can be methods based on the general theory. Papers [9] and [20] are devoted to comparison of the above two maintenance methods.

Let us start with stepwise DF $\tilde{F}(t)$ having the form:

$$\tilde{F}(t) = \begin{cases} p_0 = 0, & 0 \leq t < \Delta; \\ p_1, & \Delta \leq t < 2\Delta; \\ \vdots & \\ p_k, & \Delta k \leq t < (k+1)\Delta. \end{cases} \tag{4.3}$$

Assume that PR can be carried out only at time instants $\tau = mk + 0$, $m = 0, 1, 2, \ldots$. Define a random process with states $E_{-1}, E_0, \ldots, E_n$, where state E_k, $k \geq 0$, represents the age of the element at time $t = k\Delta$, and state E_{-1} the failure of the element (see Figure 17). The states are observed at instants

$$\tau = m\Delta + 0, \qquad m = 0, 1, 2, \ldots.$$

ξ_t is a Markov chain with the following transition probabilities:

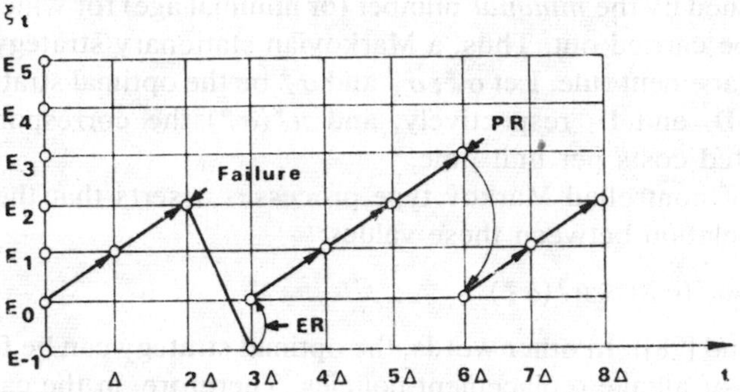

Figure 17. Sample function of the controlled process ξ_t.

$$P\{\xi_{t+1} = E_j \mid \xi_t = E_k\} = \begin{cases} \dfrac{p_{k+1} - p_k}{1 - p_k}, & j = -1; \\ \dfrac{1 - p_{k+1}}{1 - p_k}, & j = k + 1. \end{cases} \tag{4.4}$$

The control of the process ξ_t is carried out in the following manner. In every state E_k, $k \geqslant 0$, two CAs are available: to do nothing ("0") or to "shift" the element instantly into state E_0 ("1"). The CA "1" is, in fact, a PR. For the transition $E_k \to E_0$, the cost c_1 is paid. In the failure state E_{-1} only one CA is made – the momentary "shift" into state E_0 representing the ER, after which the cost $c_a = 1$ is paid. Therefore,

$$P\{\xi_{t+1} = E_j \mid \xi_t = E_{-1}\} = P\{\xi_{t+1} = E_j \mid \xi_t = E_0\}. \tag{4.5}$$

Denote by $\mathbf{D}_0$ the class of all possible control rules (strategies). Every strategy is a rule prescribing the choice between CA "0" and "1" for all states; this choice – in the general case – can be based on some "history" of the process. Let $\mathbf{D}_2 \subset \mathbf{D}$ denote the class of strategies when the control rule is as follows: take CA "1" if and only if time interval $N\Delta$ has elapsed since the last CA "1". In fact, $\mathbf{D}_2$ is the class of all block replacement policies; ΔN is a period of block replacement. Together with $\mathbf{D}_2$ consider a subclass $\mathbf{D}_1$ of all Markovian stationary strategies. Note that "Markovian" means that the choice of CA must be a function only of the actual observed state; "stationary" means that this function does not depend on the number of the step. In other words, a member of the class $\mathbf{D}_1$ has the following form: for a subset $\mathbf{E}_1$ of states, the CA "0" is prescribed; for the complementary subset, the CA "1" has to be selected. But the process ξ_t is a monotonous one and therefore, a rule which is a member of $\mathbf{D}_1$ is completely defined by the *minimal* number (or minimal age) for which the CA "1" must be carried out. Thus, a Markovian stationary strategy is a certain age replacement rule. Let σ_0^*, σ_1^* and σ_2^* be the optimal strategies in classes $\mathbf{D}_0$, $\mathbf{D}_1$ and $\mathbf{D}_2$ respectively, and $u_i^*(\sigma_i^*)$ the corresponding minimal expected costs per unit time.

The theory of controlled Markov type processes asserts that there is the following relation between these values:

$$u_0^*(\sigma_0^*) = u_1^*(\sigma_1^*) \leq u_2^*(\sigma_2^*) \tag{4.6}$$

(see [29], [77] and [78]). In other words, the optimal strategy can be found in the subclass of all age replacement policies. Therefore, in the case of stepwise functions $\tilde{F}(t)$

$$\min_{T>0} u_1(T) \leq \min_{T>0} u_2(T). \tag{4.7}$$

It remains only to extend this fact to an arbitrary DF $F(t)$. It can be assumed that the minimum in (4.1) and (4.2) is attained in an interval (t_1, ∞), $t_1 > 0$. Denote by $u_i(T, \Delta)$ the cost rates in (4.1) and (4.2) if the DF $F(t)$ is a stepwise function whose values in the points $t = k\Delta$, $k = 0, 1, 2, \ldots$, are equal to the values of the DF $F(t)$.

Using continuity arguments, it is possible to prove that for every $\epsilon > 0$ there is such a $\Delta_0 > 0$ that for all $T \geq t_1$

$$|u_i(T) - u_i(T, \Delta)| < \epsilon, \qquad i = 1, 2. \tag{4.8}$$

if $\Delta < \Delta_0$. Fix $\epsilon > 0$. Then by (4.8)

$$u_i(T) - \epsilon \leq u_i(T, \Delta) \leq u_i(T) + \epsilon. \tag{4.9}$$

Denote

$$\begin{aligned} \inf_{T\geq t_1} u_i(T) &= u_i^*(T_i^*), \quad i = 1, 2 \\ \inf_{T\geq t_1} u_i(T, \Delta) &= u_i^*(T, \Delta), \quad i = 1, 2. \end{aligned} \tag{4.10}$$

From (4.9, 4.10) it follows that

$$\begin{aligned} u_1^*(T_1^*) - \epsilon &\leq u_1^*(T, \Delta) \leq u_1^*(T_1^*) + \epsilon, \\ u_2^*(T_2^*) - \epsilon &\leq u_2^*(T, \Delta) \leq u_2^*(T_2^*) + \epsilon. \end{aligned} \tag{4.11}$$

Using (4.7), we obtain

$$u_1^*(T_1^*) - \epsilon \leq u_2^*(T_2^*) + \epsilon. \tag{4.12}$$

Since $\epsilon > 0$ is an arbitrary value, we obtain

$$u_1^*(T_1^*) \leqslant u_2^*(T_2^*). \tag{4.13}$$

This completes the proof of the following statement: If the costs for ER and PR are the same, then block replacement cannot be more profitable than age replacement.

Table 8 presents some data concerning the numerical comparison of

Table 8.

	$k=2$			$k=4$			$k=8$		
	c_1			c_1			c_1		
	0.2	0.3	0.5	0.2	0.3	0.5	0.2	0.3	0.5
min u_1	0.906	0.979	1	0.653	0.808	0.971	0.483	0.645	0.879
min u_2	1	1	1	0.700	0.907	1	0.505	0.694	1

these two kinds of preventive maintenance for different cost values c_1 and for different gamma distributions with $k = 2, 4, 8$; $\mu = 1$.

We have formerly mentioned that in practice the PR cost in block replacement is less than in age replacement. Does the reduction in the PR cost entail a preference for the block replacement? The answer for the family of gamma distributions is given in Figure 18. The zones above the curves correspond to the case when

$$\min_{\{T>0\}} u_1(T) > \min_{\{T>0\}} u_2(T).$$

5. *Optimal Age Replacement Using only Data on Mathematical Expectation and Variance*

In the previous sections of this chapter we have considered models of preventive maintenance, assuming that the DF $F(t)$ of failure-free operation time was known. Unfortunately, this supposition is not valid in solving practical problems. It happens often, especially in the initial phase of a system's operation, that the repair group does not know the "exact" lifetime DFs and has to take decisions based on incomplete information.

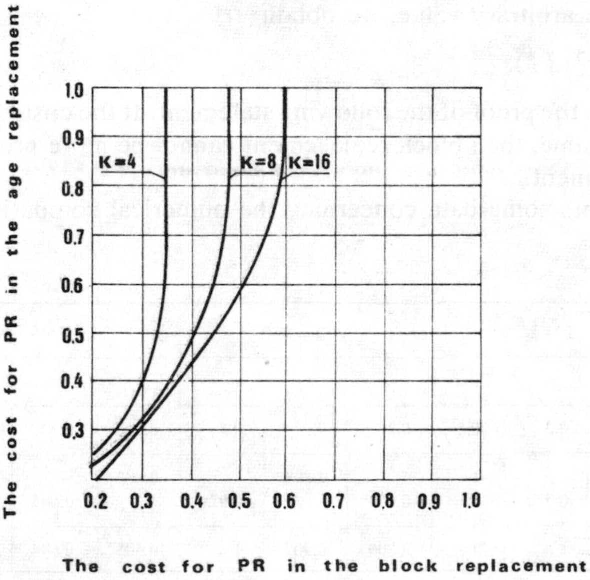

Figure 18. Comparison of block and age replacement.

We deal in this section with the following situation. The assumed knowns are the ME μ of the lifetime, its variance σ^2 and the fact that the DF of the element (system) under consideration has an increasing failure rate.

We shall restrict ourselves to the most simple model – age replacement for a single-element system (see Section 2, special case A, and Section 3, $n = 1$).

The formula for the operational readiness (3.16) can be rewritten in the following form:

$$g(T) = \left[1 + \frac{1-(1-b)(1-F(T))}{\int_0^T (1-F(x))\,\mathrm{d}x} t_a \right]^{-1} \tag{5.1}$$

where t_a is the ME of ER duration; $b = t_p/t_a$; t_p is the ME of PR duration. Introducing a new time variable $T_1 = T/\mu$, we rewrite (5.1) as follows:

$$g(T_1\mu) = g(T_1) = \left[1 + \frac{1-(1-b)(1-\hat{F}(T_1))}{\int_0^{T_1} (1-\hat{F}(y))\,\mathrm{d}y} \frac{t_a}{\mu} \right]^{-1}. \tag{5.2}$$

where $\hat{F}(y) = F(\mu y)$. Note that $\hat{F}(t)$ is a normed DF always having the ME $\hat{\mu} = 1$ and variance $\hat{\sigma}^2 = \sigma^2/\mu^2$. The problem of finding the optimal age replacement is equivalent to finding the point T_1 in which the following expression attains its minimum:

$$\varphi(T_1) = \frac{1-(1-b)(1-\hat{F}(T_1))}{\int_0^{T_1} (1-\hat{F}(y))\,dy}. \tag{5.3}$$

If $\min_{T_1>0} \varphi(T_1) = \varphi(T_1^*)$, then the optimal $T = T^* = T_1^* \mu$.

Let us consider a class $\mathfrak{F}_{1,\hat{\sigma}}$ of all such DFs which have the ME $\mu = 1$ and fixed variance $\hat{\sigma}^2$, $F(t) = 0$ for $t \leq 0$. It turns out that for members of this class $F \in \mathfrak{F}_{1,\hat{\sigma}}$, an exact upper and lower bound can be obtained. The theory of this question was elaborated on in detail by Barlow, Proshan and Marshall [10] and [11]. Paper [7] gives special tables for these bounds. It is interesting to note that the upper and lower bounds are close enough to each other. That means that the information about ME, variance and the increasing failure rate pretty well determines the lifetime DF. The reader will be convinced of this fact after acquaintance with the tables on pages 877 and 879 in [7].

Let $[1-\hat{F}(t)]^*$ and $[1-\hat{F}(t)]_*$ be the upper and lower bounds for the function $[1-\hat{F}(t)]$. Denote by $\varphi_{\max}(T_1)$ the value of $\varphi_1(T_1)$ after substituting into (5.3) $[1-\hat{F}(t)_*]$ in place of $[1-\hat{F}(t)]$. In a similar way, define $\varphi_{\min}(T_1)$. It is clear that

$$\varphi_{\min}(T_1) < \varphi(T_1) < \varphi_{\max}(T_1)\,. \tag{5.4}$$

It should be noted that this estimation of $\varphi(T)$ is not the best possible one and might be improved if the upper and lower bounds will be sought for the whole expression of $\varphi(T)$ in the class $\mathfrak{F}_{1,\hat{\sigma}}$. The method for calculating these bounds is given in [11, Theorem 4.4].

Figure 19 shows the curves of $\varphi_{\min}$ and $\varphi_{\max}$ and the curves of

$$g_{\max}(T_1) = \left[1 + \varphi_{\min}(T_1)\frac{t_a}{\mu}\right]^{-1},$$
$$g_{\min}(T_1) = \left[1 + \varphi_{\max}(T_1)\frac{t_a}{\mu}\right]^{-1}. \tag{5.5}$$

$(b = 0.1;\ t_a/\mu = 0.2;\ \hat{\sigma}^2 = 0.2)$.

By selecting the best age replacement, we shall follow the principle termed as minimax: take "the worst" value of $g(T)$ in the class $\mathfrak{F}_{1,\hat{\sigma}}$ and

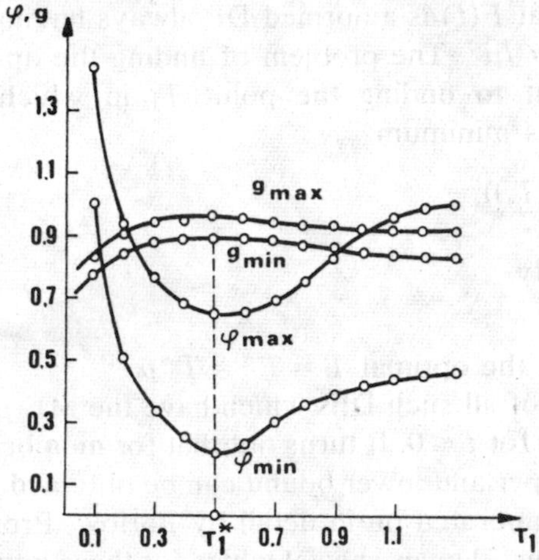

Figure 19. Dependence of φ_{min}, φ_{max}, g_{min} and g_{max} on $T_1 = T/\mu$.

choose for it the best possible value of T_1. In our case that means a choice of a preventive maintenance age T_1^* which minimizes the function $\varphi_{max}(T_1)$.

Thus, for the case plotted in Figure 19, we have to select $T_1^* = 0.5$, $(T^* = 0.5\mu)$. Here $\varphi_{max}(0.5) = 0.65$ and $\varphi_{min}(0.5) = 0.2$. This guarantees the following inequality for the operational readiness:

$$\frac{1}{1+0.2\varphi_{max}(T_1^*)} = 0.88 < g < \frac{1}{1+0.2\varphi_{min}(T_1^*)} = 0.96.$$

Table 9 contains data on the optimal "minimax" preventive maintenance period T_1^* and the values $\varphi_{min}(T_1^*)$ and $\varphi_{max}(T_1^*)$ which correspond to that period, for different values of b and $\hat{\sigma}^2 = 0.10$, 0.15, 0.20, 0.25.

In order to compare the above method with the exact one (when the lifetime DF is known), we shall consider the following example.

Let the element have a lifetime DF for which are known the ME $\mu = 2.0$ and the variance $\sigma^2 = 0.80$; the times for ER and PR are $t_a = 0.4$, $t_p = 0.04$. Define the minimax preventive maintenance period T^* and estimate the upper and lower bounds for operational readiness.

The DF $\hat{F}(t) = F(\mu t)$ has the ME $\hat{\mu} = 1$ and the variance $\hat{\sigma}^2 = \sigma^2/\mu^2 = 0.20$; $t_a/\mu = 0.2$; $t_p/t_a = b = 0.1$. For $b = 0.1$, $\hat{\sigma}^2 = 0.20$, we find in Table 9, that $T_1^* = 0.5$, $\varphi_{max} = 0.65$, $\varphi_{min} = 0.20$; the optimal age is $T^* = 0.5\mu = 1$. The bounds for OR are $0.88 < g < 0.96$.

Therefore, the value $T^* = 1$ provides the OR within limits of $0.88 \div$

Table 9
Parameters of the optimal "minimax" preventive maintenance.

	$\hat{\sigma}^2 = 0.10$		$\hat{\sigma}^2 = 0.15$		$\hat{\sigma}^2 = 0.20$		$\hat{\sigma}^2 = 0.25$	
$b = t_p/t_a$	$\varphi_{max}(T_1^*)$ $\varphi_{min}(T_1^*)$	T_1^*	$\varphi_{max}(T_1^*)$ $\varphi_{min}(T_1^*)$	T_1^*	$\varphi_{max}(T_1^*)$ $\varphi_{min}(T_1^*)$	T_1^*	$\varphi_{max}(T_1^*)$ $\varphi_{min}(T_1^*)$	T_1^*
0.05	0.37 0.10	0.5	0.48 0.10	0.5	0.56 0.10	0.5	0.63 0.11	0.5
0.10	0.47 0.20	0.5	0.56 0.20	0.5	0.65 0.20	0.5	0.72 0.21	0.5
0.15	0.55 0.25	0.6	0.64 0.25	0.6	0.72 0.31	0.6	0.79 0.38	0.6
0.20	0.62 0.33	0.6	0.72 0.33	0.6	0.79 0.39	0.6	0.86 0.45	0.6
0.30	0.76 0.44	0.7	0.84 0.50	0.7	0.91 0.56	0.7	0.97 0.61	0.7

Notations: T_1^* is the optimal period of the PR. $\varphi_{min}(T_1^*)$ and $\varphi_{max}(T_1^*)$ are lower and upper bounds for $\varphi(T_1^*)$. The bounds for the OR are

$$g_{min} = [1 + (t_a/\mu)\varphi_{max}(T_1^*)]^{-1}, \qquad g_{max} = [1 + (t_a/\mu)\varphi_{min}(T_1^*)]^{-1}.$$

0.96. Now let us choose a gamma distribution $F_k(\lambda, t)$ having the same ME and variance. Remember that $\mu = k/\lambda$, $\sigma^2 = k/\lambda^2$. Substituting values, we obtain $k = \mu^2/\sigma^2 = 5$. Set $c_a = t_a/\mu = 0.2$, $c_p = t_p/\mu = 0.02$. Examining Table 2, which gives the optimal period of replacement, we find that for $k = 4$, $g = 0.921$, and for $k = 8$, $g = 0.946$. Interpolating, we obtain that for $k = 5$, $g = 0.927$. It follows that the exact value of OR is approximately equal to the arithmetical mean of the upper and lower bounds.

It is easy to see that the smaller the ratio t_a/μ, the closer are values g_{min} and g_{max}.

This method of age replacement is expedient in a stage of operation where there is a lack of statistical data concerning the lifetime DF. In time, statistical data are accumulated and the preventive maintenance policy can be revised on the basis of more complete data.

We want to conclude this section with some general remarks about preventive maintenance when the statistical data are incomplete.

The assumption that the lifetime DF is made known throughout the exposition is, in fact, a rather strong supposition. Perhaps the most vulnerable point of all mathematical theory of preventive maintenance is, at least in its present state, that it demands too much statistical knowledge. Collecting statistical data is tedious and costly. The practical specialists often prefer to omit it, assuming that it is better to accept some "nonoptimal" but intuitively reasonable solution, rather than to undertake a lot of work without being sure that it will be more effective. The reason for such scepticism to the so-called "optimal" approach is, that by the time the statistical picture becomes clear and it would finally be possible to apply "optimal" methods, the equipment will probably become obsolete.

It has happened historically that the efforts of applied mathematicians and operations researchers were directed at solving the problems in which definite strong assumptions about initial information were made. This came about, probably, because assuming complete information is available, it is easier to obtain a formal solution of the problem. The applied mathematicians in this sense are similar to a man who has lost a coin somewhere on a dark street, and is searching for it near a lamppost because it is lighter there.

There has been an increase of works published recently which consider models of preventive maintenance with incomplete statistical data. It is possible to distinguish between two directions in these works. The first is a consideration of models where it is assumed that only some special characteristics of the lifetime, such as ME, quantiles, variance, lower or upper bounds are known. This trend of study was developed in [11], [15], [16], [25] and [75]. Surveys concerning this subject were given in [14], [51] and [61]. Speaking in general terms, this approach deals with finding the best preventive maintenance policy under the "worst" assumptions within the considered class of DFs. We have dealt with the problem in this section in the same spirit.

The main characteristic feature of the second direction in preventive maintenance studies is a "learning" process carried out at the same time as the system is functioning. Such an approach is called adaptive. In the earliest stage of operation, the repair group has some prior data about lifetime, failure rates, etc. It plans the preventive measures on the basis of this meagre information and over a certain period of operation collects data about the system's probabilistic behaviour. Afterwards, the service policy is revised and improved on the basis of this new data, and so on.†

†The reader is referred to the book [51, Chapter 9] for an example of application of adaptive policies to maintenance.

It should be noted that evaluating optimal adaptive policies is a difficult mathematical problem. In the next section of this chapter we present a very simple example of this kind.

6. Problems and Exercises

6.1. Cost Rate in Age Replacement

A device having a lifetime DF $F(t)$ is maintained according to the age replacement scheme. Let T be the "critical" age. Assume that ER lasts time t_a, PR – time t_p, and every unit of operation time is associated with a cost $r(t)$ (t is the time elapsed since the last start of operation). Write the expression for the expected cost per unit time.

Answer:

$$u_\infty = \frac{\int_0^T r(x)\,\mathrm{d}F(x) + r(T)(1-F(T)) + c_p + (c_a - c_p)F(T)}{\int_0^T (1-F(u))\,\mathrm{d}u + t_a F(T) + t_p[1-F(T)]}. \tag{6.1}$$

6.2. An Optimal Computation Procedure for a Large Multistep Problem [42]

A computer solution of a large problem consists of solving k auxiliary subproblems. The solution of one such subproblem will be called a cycle, the length of which is τ_0. The output data obtained in the ith cycle serve as the input data for the $(i+1)$th cycle, $i = 1, \ldots, k-1$. In order to preserve the results of computations from randomly occurring computer malfunctions, the following method of repeated computation is used. Several subsequent cycles n are united in one group. After completing this group, a repeated computation is carried out and the results of both computations are compared. If they coincide, the group of n cycles is finished and the computation of a new one proceeds. If the results do not coincide, the last group is computed again and the last two results are compared. The coincidence takes place if and only if there were no malfunctions during two subsequent computations of one group.

The comparison of results demands time τ_1. The probability of computer malfunction during n cycles is p_n. Find out the optimal size of a group n which provides the minimal total expected time for solving the whole problem.

Solution. Define the SMP with the following states:

E_1 – the first computation of a group is carried out;

E_2 – the repeated computation of a group is carried out; the computation has started after there were malfunctions during the previous computation of this group;

E_3 – the repeated computation of a group is carried out, starting after a previous proper computation of this group.

Show that the transition matrix is equal to

$$\mathbf{P} = \begin{Vmatrix} 0 & p_n & 1-p_n \\ 0 & p_n & 1-p_n \\ 1-p_n & p_n & 0 \end{Vmatrix}, \tag{6.2}$$

(see Figure 20).

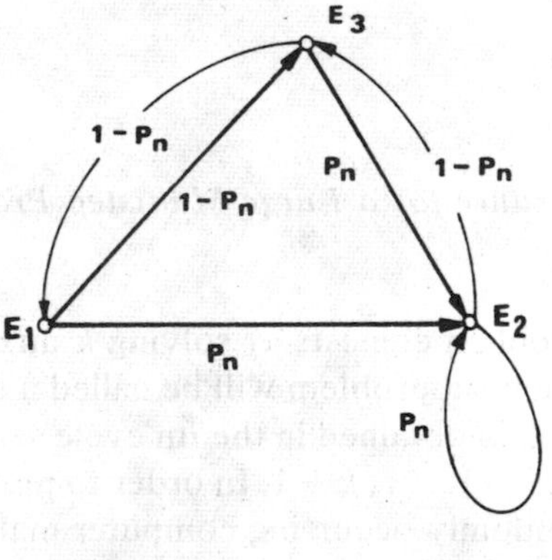

Figure 20. Transition diagram for Problem 2.

Show that the expected transition times are:

$$\mu_{12} = n\tau_0 = \mu_{13}, \qquad \mu_{31} = \mu_{23} = \mu_{22} = \mu_{32} = \tau_1 + n\tau_0. \tag{6.3}$$

The expected time spent for computing one cycle is $g_\infty = l_{11}/n$, where l_{11} is the expected return time in state E_1. Using the formula A.8 (see Appendix 2)

$$l_{11} = \frac{1}{\pi_1} \sum_{i=1}^{3} \pi_i \nu_i,$$

show that

$$g_\infty = \tau_0 + \frac{\tau_1 + n\tau_0}{n} \frac{1 + (1 - p_n) - (1 - p_n)^2}{(1 - p_n)^2}. \tag{6.4}$$

The optimal value n^* may be found from this formula if the dependence p_n of n is known. When the malfunctions have a constant failure rate, $p_n = 1 - \exp(-\lambda n \tau_0)$.

6.3. *Preventive Maintenance of a Computer*

Preventive maintenance of a computer is organized in the following manner. Planned testings are carried out after each T hours of operation. The testing lasts time t_p. If during the operation a failure occurs, its signal is received with the probability α. The elimination of the failure lasts time t_a and is carried out immediately after the signal is received. With probability $1 - \alpha$, the failure remains undiscovered up to the testing. If the ER has not been finished up to the moment of the planned testing, it is stopped and the regular testing starts (see Figure 21). The failure rate of the computer is a constant λ. Find the operational readiness.

Solution. Denote by $u(T)$ the ME of failure-free operation time in time interval $(0, T)$. It is clear that the operational readiness $g(T) = u(T)/(T + t_p)$; $u(T)$ satisfies the recurrence relation

$$u(T) = \int_0^T (1 - F(\tau))\, d\tau + \alpha \int_0^T u(T - \tau - t_a)\, dF(\tau),$$
$$u(x) = 0 \quad \text{for } x \leq 0, \qquad F(t) = 1 - \exp(-\lambda t). \tag{6.5}$$

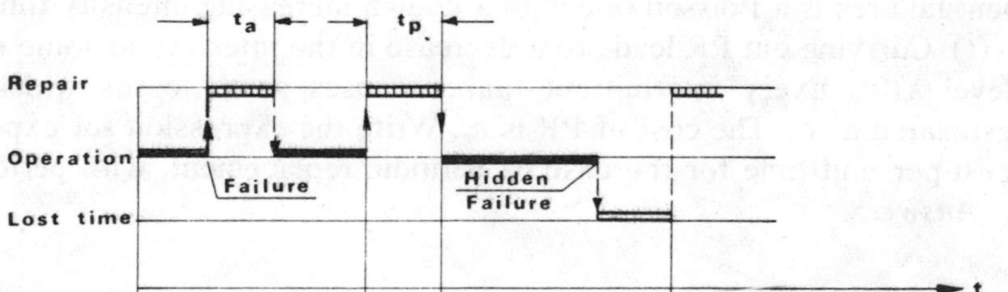

Figure 21. Scheme of the preventive maintenance of a computer (Problem 3).

6.4. Preventive Maintenance of a Unit with a Random Delay of Failure Signalling

Let $F(t)$ be the DF of failure-free operation time of a unit. If the failure occurs at instant τ, the repairman receives the information about it at time $\tau+\tau_1$, where τ_1 is a random variable with DF $G(t)$. τ_1 is independent of τ. The time limit T is set in advance, with a unit being replaced on receiving the failure signal or on reaching this limit, whichever occurs first. Every repair takes time t_0. Obtain the expression for the operational readiness.

Solution.

$$g(T)=\frac{\int_0^T (1-F(u))\,\mathrm{d}u}{\int_0^T (1-Q(u))\,\mathrm{d}u + t_0}, \tag{6.6}$$

where $Q(u)$ is the convolution of DFs $F(t)$ and $G(t)$:

$$Q(u)=\mathbf{P}\{\tau+\tau_1\leq u\}.$$

6.5. Preventive Maintenance of a Device with an Increasing Rate of Intermittent Failures

An intermittent failure lasts a short time, after which the device recovers by itself and afterwards operates reliably. The flow of intermittent failures is a Poisson one with a known increasing intensity function $\lambda(t)$. Carrying out PR leads to a decrease in the intensity to some initial level $\lambda(0)$. Every intermittent failure causes some losses which are estimated as c_a. The cost of PR is c_p. Write the expression for expected cost per unit time for the case of periodic replacement, with period T.

Answer:

$$u(T)=\frac{c_a\int_0^T \lambda(t)\,\mathrm{d}t + c_p}{T}. \tag{6.7}$$

6.6. Section 2.3. Special Case C, "False Alarm"

Assume that in addition to the statement of the problem in special case B (see Section 2.3), there is a source of a "false alarm" (FA). The FA is given according to the following scheme: if the element is in an operable state at time t, the probability of sending the FA in the interval $(t, t+\Delta t)$ is equal to $\lambda\Delta t + o(\Delta t)$. The repair group cannot distinguish between the FA and a "real" failure signal. Because of this, having received the failure signal, an inspection is carried out during time t_s^*, as a result of which the state of the device becomes clear: it has or has not failed. In the case of failure, the ER is carried out, otherwise, the PR. Write an expression for the OR for a fixed inspection period T.

Solution. Define the SMP with the states:

E_1 – element is operating;
E_2 – element is inspected, there is no failure;
E_3 – element is inspected, the inspection has started in the absence of a failure signal; there is a failure at the beginning of the inspection;
E_4 – the element is under ER;
E_5 – the element is inspected after a failure signal, there is a failure;
E_6 – the element is inspected, the inspection was initiated by the FA.

Graph of these states is shown in Figure 22. The transition probabilities are:

$$p_{12} = [1 - F(T)]\,e^{-\lambda T}, \qquad p_{13} = (1-\alpha)\int_0^T e^{-\lambda T}\,dF(t),$$
$$p_{15} = \alpha\int_0^T e^{-\lambda t}\,dF(t), \qquad p_{16} = \int_0^T \lambda\,e^{-\lambda t}[1 - F(t)]\,dt. \tag{6.8}$$

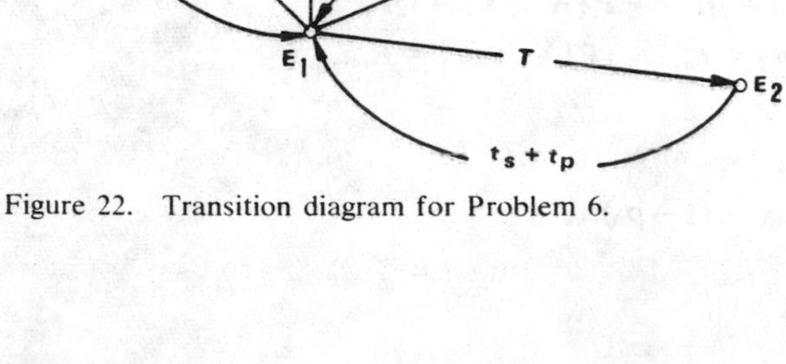

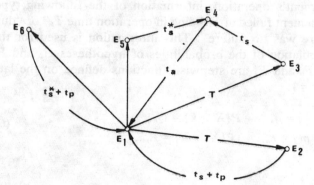

Figure 22. Transition diagram for Problem 6.

The transition times have the following DFs:

$$F_{15}(t)=\begin{cases}\int_0^t e^{-\lambda x}\,dF(x)\Big/\int_0^T e^{-\lambda t}\,dF(t), & t<T;\\ 1, & t\geq T\end{cases}$$

$$F_{16}(t)=\begin{cases}\int_0^t (1-F(x))\lambda\,e^{-\lambda x}\,dx\Big/\int_0^T [1-F(x)]\lambda\,e^{-\lambda x}\,dx, & t<T;\\ 1, & t\geq T.\end{cases}\tag{6.9}$$

$\nu_1(T)$ must be calculated according to formula (3.42) from Chapter 1. A one-step return must be assumed equal to the ME of failure-free operation time with the initial state E_1:

$$\omega_1(T)=\boldsymbol{M}\{\min(\tau,\tau_{fa},T)\},\tag{6.10}$$

where τ_{fa} is a random variable with DF $1-\exp(-\lambda t)$. The other $\omega_i(T)=0$.

6.7. Adaptive Age Replacement

This problem has to serve as an illustration of the simplest adaptive preventive maintenance models. *A priori* information concerning the element's DF is the following: probabilities p_0 and $q_0=1-p_0$ of the hypothesis H_0 and H_1 that the element's lifetime DF is either $F_0(x)$ or $F_1(x)$ (these functions are assumed to be completely known). During the time of the element's operation, information of the following type is received: "the element failed after being in operation time T_1" or "during the time T_2 there was no failure". This information is used for the *a posteriori* recalculation of the probabilities of hypotheses H_0 and H_1.

Assume that F_0 and F_1 are stepwise functions defined on the lattice with the step $\Delta=1$:

$$\begin{aligned}&F_0\colon\ \boldsymbol{P}\{X=k\}=q_k, \qquad \boldsymbol{P}\{X>k\}=Q_k;\\ &F_1\colon\ \boldsymbol{P}\{X=m\}=r_m, \qquad \boldsymbol{P}\{X>m\}=R_m.\end{aligned}\tag{6.11}$$

Denote

$$f(k,p_0)=p_0q_k+(1-p_0)r_k,$$

$$\Phi_j(p_0) = \sum_{k>j} f(k, p_0).$$

We shall now consider the functioning over the interval $[0, T]$, where $T = n\Delta$, n is an integer. PRs are allowed only at instants $\tau = k\Delta + 0$, $k = 1, \ldots, n$. Assume $\Delta = 1$.

At $\tau = 0$, a new element starts to operate. The CA is a choice of time for carrying out the PR. If the element fails, it is immediately replaced by a new one. The cost of PR is C, the cost of failure is 1, $C < 1$. At time $\tau = T + 0$, the performance is finished; the failure's penalty (if it occurs) is counted at time $\tau = T$. Our goal is to find the optimal rule for carrying out PRs so as to minimize the total expected cost on the interval $[0, T]$.

Solution. Let $A(H_0|m, p_0)$ denote the *a posteriori* probability of hypothesis H_0 calculated under assumption that a lifetime of $m\Delta$ was observed and that the *a priori* probability of H_0 was p_0. According to Bayes' formula,

$$A(H_0|m, p_0) = \boldsymbol{P}\{H_0|X = m, p_0\} = \frac{p_0 q_m}{p_0 q_m + (1 - p_0) r_m}. \tag{6.12*}$$

Similarly, $A(H_0|>l, p_0)$ denotes the *a posteriori* probability of H_0 calculated under assumption that a lifetime $\tau > l\Delta$ was observed and the *a priori* probability of H_0 was p_0:

$$A(H_0|>l, p_0) = \boldsymbol{P}\{H_0|>l, p_0\} = \frac{p_0 Q_l}{p_0 Q_l + (1 - p_0) R_l}. \tag{6.12**}$$

We agree on doing an *a posteriori* recalculation of the probability of H_0 each time when a failure is observed or when a PR is carried out.

If at a certain moment when ER or PR are made, the *a posteriori* probabilities of H_0 and H_1 are p_0^* and $(1 - p_0^*)$, then preventive maintenance is planned as if there is at our disposal an element having DF

$$F(x) = p_0^* F_0(x) + (1 - p_0^*) F_1(x).$$

Denote by $u(k; p_0)$ the minimal total expected cost for an interval $[n - k, n]$ calculated under the following conditions:

(1) a new element is put into operation at $\tau = n - k$;
(2) $\boldsymbol{P}\{H_0\} = \hat{p}$ at time $\tau = n - k$.

Let us derive a recurrence relation which will serve for finding the optimal strategy.

From the statement of the problem, it follows for an one-step operation

$$u(1;\hat{p}) = f(1,\hat{p}).$$

Now consider an interval $[n-m, n]$. A new element starts to operate at $\tau = n - m$. Let the decision to carry out the PR after k time units be made (at $\tau = n - m + k + 0$). The element will fail with the probability $f(i,\hat{p})$ at instant $\tau = n - m + i$, as a result of which cost 1 will be paid for the failure plus the cost

$$u(m-i; A(H_0|i,\hat{p}))$$

which will be accumulated in the remaining period of time. (Note that the second term in this formula represents the new *a posteriori* knowledge about H_0 based on newly received data – the element's lifetime duration was i units.)

If, on the other hand, the element does not fail up to the moment $\tau = n - m + k$ (this event has the probability $\Phi_k(\hat{p})$), then the cost paid will be

$$C + u(m-k; A(H_0|>k;\hat{p})).$$

From this reasoning follows the desired recurrence relation:

$$\begin{aligned} u(m;\hat{p}) = \min_{1\le k\le m} &\Big\{\sum_{i=1}^{k} f(i,\hat{p})[1 + u(m-i; A(H_0|i,\hat{p}))] \\ &+ \Phi_k(\hat{p})[C(1-\delta_{mk}) + u(m-k; A(H_0|>k;\hat{p}))], \end{aligned} \qquad (6.13)$$

$$u(0;\hat{p}) \equiv 0.$$

Multiplier

$$\delta_{mk} = \begin{cases} 0, & m \neq k, \\ 1, & m = k, \end{cases}$$

reflects the fact that if a PR is appointed at the end of the operation interval, then no cost is paid for it.

For better understanding of relation (6.13) we stress that the value $A(H_0|\cdot)$ represents the "state of our learning" about DFs F_0 or F_1. It plays the same role as the index of state E_i in nonadaptive maintenance models.

As to the computational aspect of the problem, it should be mentioned that at every step it is necessary to know the value of $u(k,\hat{p})$, for all values of $\hat{p}$, $0 \le \hat{p} \le 1$. Therefore, for computerized calculations of $u(m, p)$, it is necessary to have tables in the memory for $(m-1)$ one-dimensional functions. It is clear that this is a problem essentially more complicated than all those considered before. The computational difficulties would increase if we consider, for example, a model with hypotheses H_0, H_1, H_2. Then it would be necessary to have values of functions of two variables in the memory.

6.8. *Discovering a Malfunction in an Automatic Machine*

An automatic machine produces one article every unit of time. If it is in a good state, then there is the probability p_0 that the produced article is defective. If the machine has a malfunction (failure) this probability is equal to p_1, $p_1 > p_0$. The entire time of machine operation is divided into periods, during which N articles are produced. It is assumed that the malfunction can arise with constant probability γ at the beginning of each period. The state of the automatic machine is tested by taking samples of n articles at the end of each period. All articles in the samples are examined and if the number $\boldsymbol{X}$ of the defective ones is equal to k or more (k is the prescribed "critical level", $k \leq n$), it is assumed that there is a malfunction. In this case, the machine is down for a time m_p, as a result of which it becomes clear if there is a malfunction or not. If the alarm is "false", the machine starts to operate immediately. If not, an additional time m_a is spent for carrying out the ER. After the machine is repaired, it has the initial level p_0 of producing defective pieces.

The problem is to find the optimal value k which provides the maximal expected number of nondefective articles produced in a unit of time. This problem is a slight generalization of the one given in [57].

Solution. Define a SMP with the states:

E_1 – the machine is operating, no malfunctions;
E_2 – the machine is operating, there is a malfunction;
E_3 – the machine is down, it was stopped after the malfunction appeared;
E_4 – the machine is down, it was stopped because of a "false alarm".

Transition probabilities $p_{31} = p_{23} = p_{41} = 1$. Let $b_0(n, k)$ denote the probability of finding in a sample k or more defective units, if the machine is in order, and let $b_1(n, k)$ be the same probability for a machine having a malfunction. It is clear that

$$
\begin{aligned}
b_0(k, n) &= \boldsymbol{P}\{\boldsymbol{X} \geq k; p_0\} = \sum_{i=k}^{n} \binom{n}{i} p_0^i (1 - p_0)^{n-i} \\
b_1(k, n) &= \boldsymbol{P}\{\boldsymbol{X} \geq k; p_1\} = \sum_{i=k}^{n} \binom{n}{i} p_1^i (1 - p_1)^{n-i}.
\end{aligned}
\tag{6.14}
$$

Let Y denote the number of the first period at the beginning of which the malfunction occurs. According to the statement of the problem

$$P\{Y=l\}=(1-\gamma)^{l-1}\gamma, \qquad l\geq 1,$$
$$P\{Y>m\}=\sum_{l=m+1}^{\infty}P\{Y=l\}=(1-\gamma)^m. \tag{6.15}$$

The transition $E_1\to E_4$ occurs if and only if the alarm signal was sent before the malfunction appeared. The probability that it happens at the end of mth period is equal, through (6.14, 6.15), to

$$p_{14}(m)=[1-b_0(k,n)]^{m-1}b_0(k,n)(1-\gamma)^m.$$

From this it follows

$$p_{14}=\sum_{m=1}^{\infty}p_{14}(m)=b_0(k,n)(1-\gamma)[1-(1-b_0(k,n))(1-\gamma)]^{-1}. \tag{6.16}$$

Later on, $p_{12}=1-p_{14}$. Now it is easy to calculate the limiting probabilities π_i of states E_i. They are:

$$\pi_1=(3-p_{14})^{-1}, \qquad \pi_2=(1-p_{14})(3-p_{14})^{-1}=\pi_3,$$
$$\pi_4=p_{14}(3-p_{14})^{-1}. \tag{6.17}$$

Now calculate the holding times. It is clear that $\nu_3=m_p+m_a$, $\nu_4=m_p$. The machine remains a random number of periods Z, $Z>m$, in state E_1 if and only if there were not both malfunctions and "false alarms" during the first m periods. Obviously,

$$P\{Z>m\}=(1-b_0(k,n))^m(1-\gamma)^m, \qquad m\geq 1, \tag{6.18}$$

and therefore,

$$\nu_1=NM\{Z\}=N\sum_{k=1}^{\infty}P\{Z=k\}k=N\left[1+\sum_{r=1}^{\infty}P\{Z>r\}\right],$$
$$=N[1-(1-\gamma)(1-b_0(k,n))]^{-1}. \tag{6.19}$$

The mean holding time in E_2 is equal to $NM\{w\}$ where w is a random number of periods which elapsed from the moment of the malfunction up to its discovery. It is easy to check that

$$\nu_2=N/b_1(n,k). \tag{6.20}$$

Let us define the returns for states E_i as a mean number of nondefective articles produced in the corresponding state. Obviously,

$$\omega_1 = \nu_1(1-p_0), \qquad \omega_2 = \nu_2(1-p_1), \qquad \omega_3 = \omega_4 = 0. \tag{6.21}$$

Substituting (6.17)–(6.21) into the formula $g = \Sigma_{i=1}^4 \pi_i\omega_i / \Sigma_{i=1}^4 \pi_i\nu_i$, we obtain, after some manipulations, the desired expression

$$g(k) = \frac{(1-p_0) + (1-p_1)b_1^{-1}(n,k)\gamma}{1 + (m_a^1 + b_1^{-1}(k,n))\gamma + m_p^1(\gamma + b_0(n,k) - \gamma b_0(n,k))}, \tag{6.22}$$

where $m_a^1 = m_a/N$, $m_p^1 = m_p/N$. Investigating this expression, we can find the optimal k^* providing the maximal value of $g(k)$.

Chapter 3

PREVENTIVE MAINTENANCE BASED ON SYSTEM PARAMETER SUPERVISION

1. Introductory Remarks

In the previous chapter we dealt with the preventive maintenance of objects for which only two states could be distinguished: failure and nonfailure. All *a priori* known information about the maintained system was presented in its failure-free time DF. The only *a posteriori* received data were the time elapsed since the last start of operation (system's "age") and sometimes also the signal that the failure had taken place. For a model of failures in many technical systems, it is typical that the failure is preceded by a certain kind of accumulation of damages (deterioration). We mention failures such as those caused by the gradual drift of the system's output characteristics beyond permissible regions (radio and electronic equipment), wear and tear fatigue (mechanical devices), etc. In reliability literature such failures are called "gradual" [68, Chapter 1]. It should be noted that behind the so-called "instantaneous" failures that lead to a sudden loss of ability to function, there exists in many (if not all) cases a nonobservable hidden process of accumulation of damages (see discussion in [41, Chapter 4]). The general scheme in which these failures appear involves gradual changes in the physical properties of the system which inevitably lead to its weakening; as a result, interaction with outside impacts causes sudden failure in objects whose "resistance" ability has been decreased. As a typical example, we cite the breakdown of machine parts occurring under the influence of a peak load when they have been weakened as a result of corrosion or through wear and tear.

Several studies were devoted to discovering and examining parameters which were good enough to forecast the objects' failure. In reliability literature, data have been published about the increase of thermal noise in electronic tubes preceding the tube failure [59]. Those authors who studied the failures of relay contacts mention the fact that failure can be predicted on a basis of measuring the "noise" on the contact surfaces. Thus, stating it more formally, it is possible to discover in the system one or several randomly changing parameters which provide information about the "inside" state of the object and the possible appearance of a failure. Unfortunately, in practical situations we can exploit the existence of such "prognostic" parameters only in rare cases. One reason is the difficulty in observing and measuring these parameters. For example, it is very difficult to measure fatigue damages; the measurement of some parameters can be carried out only by dismantling the device. The other reason is that we do not know the stochastic relation between the value of certain prognostic parameters and the remaining random time-to-failure. In the above situations, it remains only to propose some reasonable failure model, to try to obtain the lifetime DF, estimate its parameters and organize the preventive maintenance only on the basis of observing the system's age or the accrued operation time.

Our approach to preventive maintenance in this chapter will be based on more optimistic suppositions. We suppose that there exists a possibility to supervise system's "inside" state and utilize these data for organizing preventive maintenance. Accordingly, the following assumption will be made:

(a) The "inside" ("interior") state of the system can be characterized by some randomly changing time-depending parameter $\eta(t)$.

(b) A random event called "failure" is defined. It is supposed that the failure is either a certain random event whose probability of occurrence depends on the value of the parameter $\eta(t)$, or is defined as the parameter $\eta(t)$ entering some "critical" domain. In the last case it is possible to determine several "intermediate" state of the system, apart from states "the system is new" and "the system has failed".

(c) The parameter $\eta(t)$ is accessible for continuous or discrete observations.

(d) The probabilistic laws governing the changes of the parameter $\eta(t)$ are known. If this parameter is stochastically connected with the failure, then it is supposed that we know the stochastic relation between the parameter $\eta(t)$ and the probability of failure appearance.

Example C in Chapter 1, Section 1 deals with preventive maintenance for the above assumptions. It is clear that supervised preventive maintenance based on the observation of parameters must be more effective than preventive maintenance based only on the system's DF and age. This is shown by the fact that no unnecessary repairs are carried out when the failure risk is small and the parameter's $\eta(t)$ value is far from the "dangerous" zone. Probably, the most suitable objects for making supervised preventive maintenance, using prognostic parameters, are electronic devices. Usually every such device (amplifier, generator, pulse former, etc.) can be examined relatively easily in order to determine the values of its main output characteristics. In this sense the electronic device is preferred to a mechanical one, because measuring parameters in a mechanical device can be carried out only after dismantling, or by especially constructed flaw detectors.

In practice, suppositions a, b, c, hold and are not restrictive. The key supposition is the last one (d). Study of the statistical behaviour of a random process is not an easy task. Here a contradiction arises which is typical for constructing of probabilistic models. On one hand, while striving for an exact model which will reflect "all" of the features of the real phenomena, we are inevitably forced to incorporate into our model a large number of parameters, all of which must be later estimated from empirical data. However, it tends to nullify the precision of our "complete" model. This danger also exists in our case. It may happen that formal preventive maintenance models which use data about the system's randomly changing parameters will be practically useless or, at the very best, no better than simpler age or block replacement models. The reason is that the more complicated models involve much statistical data which could never be provided with sufficient exactness.

Nevertheless, there is a group of technical systems where the situation is less pessimistic and where we can hope that supervised preventive maintenance policies will not be left "hanging in the air". We refer to systems having standby redundance. In these systems the "state" can be characterized naturally as a number of failed standby parts. In many cases, the exponential DF is valid for a standby part. This means, in turn, that the evolution of a system having standby parts can be described by means of a Markov-type random process with a finite number of states. From the formal point of view, this is a property of crucial importance. The practical applicability of the theory is provided by the fact that in this case only the most simple statistical data, i.e. the failure rates of standby parts, are involved.

Throughout this chapter it will be assumed that $\eta(t)$ is a Markov process having states $E_0, E_1, \ldots, E_n$.

A brief discussion of the chapter contents follows. The next section deals with a system having one standby part. In a formal sense, this model refers to a three state controlled SMP considered on a finite time interval. In the framework of this scheme, a special additional constraint on the inspection period is introduced which guarantees the prescribed nonfailure probability between inspections.

Section 3 deals with a partially repaired system which functions up to the moment of the first failure. The system has one "critical" state where a repair is done. Formally speaking, this is an example of the controlled SMP with an absorbing state (failure of the system).

Section 4 describes a situation which is specific for a system with "cold" standby: several inspection times are foreseen when it is possible to decide how many standby parts must be engaged for one time period up to the next inspection. The goal is to minimize the probability of system's failure. Like Section 2, the formal model refers to a finite-step controlled process.

In Section 5 a "multiline" system consisting of n identical subsystems (lines) is considered. Its characteristic feature is that it can function when k $(1 \leq k < n)$ "lines" have failed; when the failed "lines" are being repaired, the whole system must be idle. The prototype for a multiline system is a special kind of automatic production lines. The goal is to find an optimal service rule (the "critical" number of failed lines) which provides the maximal efficiency of the whole system.

Section 6 introduces a somewhat different system failure scheme: the parameter $\eta(t)$ available for observation only predicts the failure of the system: the probability of the failure depends on the value of this parameter (which changes randomly). The preventive maintenance scheme is based on observation of this prognostic parameter; preventive actions are made when the prognostic parameter enters some critical region.

Section 7 discusses a quite general maintenance scheme: the set of available CAs contains arbitrary shifts from a "dangerous" state E_i to a "safe" state $E_{\varphi(i)}$ and an appointment of the next inspection time $T_{\varphi(i)}$. The process is considered on an infinite time interval and the goal is to find an optimal stationary strategy providing the minimal expected cost per unit time.

Section 8 introduces a new situation in the control of a SMP: a presence of one or several linear constraints on state probabilities. For example, one could look for the optimal strategy minimizing the expected cost with the requirement that the frequency of entering a given state must be lower than a certain value. An interesting feature of this model is that in contrast to the ordinary situation, here randomized strategies sometimes are optimal.

In practice, often "threshold" or control-limit policies are very popular. In this case all states of the system are divided into two groups $\{E_1, \ldots, E_k\}$

and $\{E_{k+1}, \ldots, E_n\}$; if the system enters state E_r, $r \geqslant k$, it is immediately shifted to state E_1. Nothing is done when the system enters states $E_1, \ldots, E_k$. Section 9 investigates some situations when this kind of control turns out to be optimal in respect to the average cost per unit time.

2. Optimal Preventive Maintenance for System with Standby Functioning on a Finite Time Interval

2.1. Statement of the Problem

For the sake of simplicity, the problem will be expounded in application to a very simple system having only one main and one standby unit (see Figure 23).

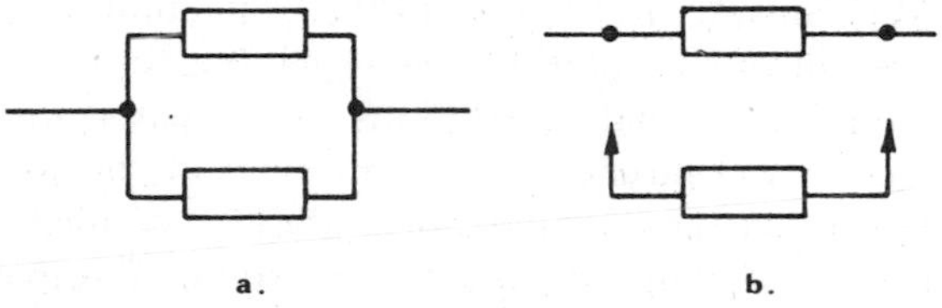

Figure 23. A system with one main and one standby unit: (a) "hot" standby; (b) "cold" standby.

We assume that both units have a constant failure rate λ. For our consideration it is not important how the standby unit is switched on; it might be "hot" or "cold" standby. The system must work a prescribed time θ. It is also assumed that the system is located in a hard-to-reach place and because of that, there is no information about failure. To maintain the system in a good state, inspections are made from time to time. The repair group checks the state of the system and makes a repair if it is expedient. If during the inspection, it is ascertained that only one unit has failed, then there is the possibility of replacing the failed unit with a new one, or postponing the repair until the next inspection. If the inspection reveals the failure of all systems, i.e. two units have failed, there are then also two possibilities concerning the repair: replacing two units or only just one. After the inspection and switching the system into operation, a new inspection time is fixed. This must be done in such a way that the probability of a failure up to the next inspection does not exceed a given constant γ. The maximal interval between two subsequent inspections can be found as follows. Let us assume, for example, that after the inspection and the repair, the system started to work with two good units. Let $F_0(t)$ denote the DF of the failure-free operation time for this system. We fix the period of inspection T_0 so that

$$F_0(T_0) \leq \gamma. \tag{2.1}$$

It is easy to verify that for the "hot" standby, $F_0(t) = (1 - e^{-\lambda t})^2$ and in the case of the "cold" standby, $F_0(t) = 1 - e^{-\lambda t}(1 + \lambda t)$. If after an inspection, the system starts to work with only one good element, the period to the next inspection T_1 must satisfy the condition

$$F_1(T_1) \leq \gamma, \tag{2.2}$$

where $F_1(t) = 1 - e^{-\lambda t}$ is the DF of the time-to-failure for one unit.

It is assumed that the penalty cost paid for the failure occurring between subsequent inspections is equal to a_f. The cost of the inspection is equal to a_s. The cost for the repair of two units is a_{2p}, the repair cost of one unit is equal to a_p. It is assumed also that inspections and repairs are done instantaneously and that at the moment $t = \theta$ (the end of the job), the system is checked and the failure cost is paid if the failure is revealed.

We shall consider the problem of finding the optimal preventive maintenance policy in respect to minimizing the average expenditures over a given time interval $[0, \theta]$. Such a statement of the problem is expedient if the time for servicing is much less than the time the system has to work. In the case where the time spent for the service is not negligible, another statement of the problem seems to be more realistic: find a replacement policy which maximizes the average time that the system is in working order. One example of that kind is given in Section 10, Problem 1.

We point out the principal difference between this problem and the ones considered earlier. In this case, there is, besides the state "two units are good" and "two units failed", one intermediate state "exactly one unit has failed". This radically changes the situation and makes it possible to effect not only a total repair of the system, but also a "partial" repair, where only one unit is replaced by a new one, after the failure is revealed.

2.2. The Observed Process ξ_t, Service Strategy, and Costs

Three states will be defined in the system: E_0 – two units are good;

E_1 – one unit has failed; E_2^* – failure of the system (two units have failed). To make the exposition simpler, the time scale will be assumed to be a discrete one. The interval $[0, \theta]$ will be divided into N small subintervals. Each of them has the length $\Delta: \theta = N\Delta$, $\Delta = 1$. It is assumed also that the periods T_0 and T_1 are multiples of $\Delta: T_0 = K_0\Delta$, $T_1 = K_1\Delta$. Because $\Delta = 1$, we shall count off the time in integers $0, 1, 2, \ldots, k, \ldots, N$.

Let us assume that at time k, an inspection took place and the state E_i, $i = 1, 2$, is discovered. The control action $\sigma_i(k)$ for that state is the "shift" into state E_j, $j \leqslant i$, and the choice of the next inspection time which depends both on E_j and k. The dependence here is very simple: if $N - k > K_j$, that is, the time up to the end of the job is greater than the value $T_j = K_j$, the next inspection will take place at time $k + K_j$, otherwise at time $\theta = N$. Control actions for states E_2 and E_1 might be written in the following form:

$$\begin{aligned} \sigma_2(k) &= (E_{\varphi(2,k)}; \min(N-k; K_{\varphi(2,k)})), \\ \sigma_1(k) &= (E_{\varphi(1,k)}; \min(N-k; K_{\varphi(1,k)})). \end{aligned} \tag{2.3}$$

Here $\varphi(i,k)$ is equal to 0 or 1. If $\varphi(2,k) = 1$, the first row of (2.3) means the following: if state E_2^* is observed, the system must be transferred into state E_1 (only one unit is replaced) and the period $\min(N-k; K_1)$ is fixed for the next inspection. For state E_0 there is no choice and therefore the CA is written in the form

$$\sigma_0(k) = (E_0; \min(N-k; K_0)).$$

The vector

$$\boldsymbol{\sigma}(k) = \{\sigma_i(k), i = 0, 1, 2\} \tag{2.4}$$

defined for $k = 0, \ldots, N-1$ is called the service strategy. Let us assume that E_{i_0} is the state of the system at moment $t_0 = 0$ and that strategy (2.4) is chosen. Then it defines the time $\mathbf{t}_1$ of the next inspection and the "stochastic mechanism" for the appearance of state E_{i_1} at time $\mathbf{t}_1$. For the exact description of this mechanism, some new notations must be given.

Let us define a random process ξ_t called an *observed* process over the interval $[0, \mathbf{t}_1]$:

$$\begin{aligned} &\xi_0 = E_{i_0}, \qquad \text{for } 0 \leqslant t < \mathbf{t}_1, \quad \xi_t = E_{i_0}; \\ &\xi_{\mathbf{t}_1} = E_{i_1}. \end{aligned}$$

In other words, the process ξ_t takes the value of the observed state of the system at moment $t_0 = 0$ and keeps it until the moment of the next observation. If the next observed state at moment t_1 is E_{i_1}, the next value of the process ξ_t will be according to its definition E_{i_1}, and so on.

Let $P_{i,j}(t)$ denote the probability that at the end of the interval $(0, t)$ in the system there will be j failed units, if it is assumed that at moment $t = 0$ the number of failed units is equal to i and that during the entire period $(0, t)$ the system was left "to its own devices".

Now it is possible to write the expression for the probability of observing state E_j at moment t_1 if state E_1 was observed at moment t_0:

$$P\{\xi_{t_1} = E_j \mid \xi_{t_0} = E_i\,;\ \sigma_i(0)\} = P_{\varphi(i,0),j}(\min(N, K_{\varphi(i,0)})). \tag{2.5}$$

In fact, at $t_0 = 0$ the system is transferred into state $E_{\varphi(i,0)}$; the next inspection will take place after the time $x = \min(N\,;\ K_{\varphi(i,0)})$; therefore, the probability to observe state E_j at the next inspection is equal to $P_{\varphi(i,0),j}(x)$.

In an analogous way, one might describe the stochastic mechanism of the observation of state E_{i_2} at the second inspection time, and so on. We shall describe this mechanism in a general form. Let state E_{i_0} be fixed at time $t_0 = 0$ and strategy $\boldsymbol{\sigma}(k)$ be chosen. Let $\{t_n\}$ be the sequence of time moments when the system is observed, $n = 0, 1, \ldots$, and E_{i_n} is the state observed at moment t_n.

The observed process ξ_t is defined in the following manner:

$$\xi_t = E_{i_n}, \qquad \text{for } t_n \leqslant t < t_{n+1}, \quad n \geqslant 0. \tag{2.6}$$

The transition probabilities for the process ξ_t have the form:

$$P\{\xi_{t_{n+1}} = E_j \mid \xi_{t_n} = E_i\,;\ \sigma_i(t_n)\} = P_{\varphi(i,t_n),j}(K^*_{\varphi(i,t_n)}), \tag{2.7}$$

where $K^*_{\varphi(i,t_n)} = \min(N - t_n\,;\ K_{\varphi(i,t_n)})$. We give an example illustrating the definition of process ξ_t.

Example. The strategy has the following form: in state E_2^*, the complete repair is always performed (the system is transferred into state E_0). If at the time of an inspection, state E_1 is detected, no repair is made. After detecting state E_0 and after the repair $E_2^* \to E_0$, the next inspection is planned after $K_0 = 8$ time units. The length of the functioning period is $N = 40$ time units. At the end of this period an inspection is also planned.

In Figure 24a, a record of the service process is drawn. In the lower part of Figure 24, the sample function of the corresponding observed process ξ_t is plotted. According to the above, the strategy $\boldsymbol{\sigma}_{(k)}$ can be written in the following form

$$\begin{aligned}\boldsymbol{\sigma}(k) = \{&E_0, \min(40-k, 8); \quad E_1, \min(40-k, 5);\\ &E_0, \min(40-k, 8)\}, \quad k = 0, 1, \ldots, 39.\end{aligned} \tag{2.8}$$

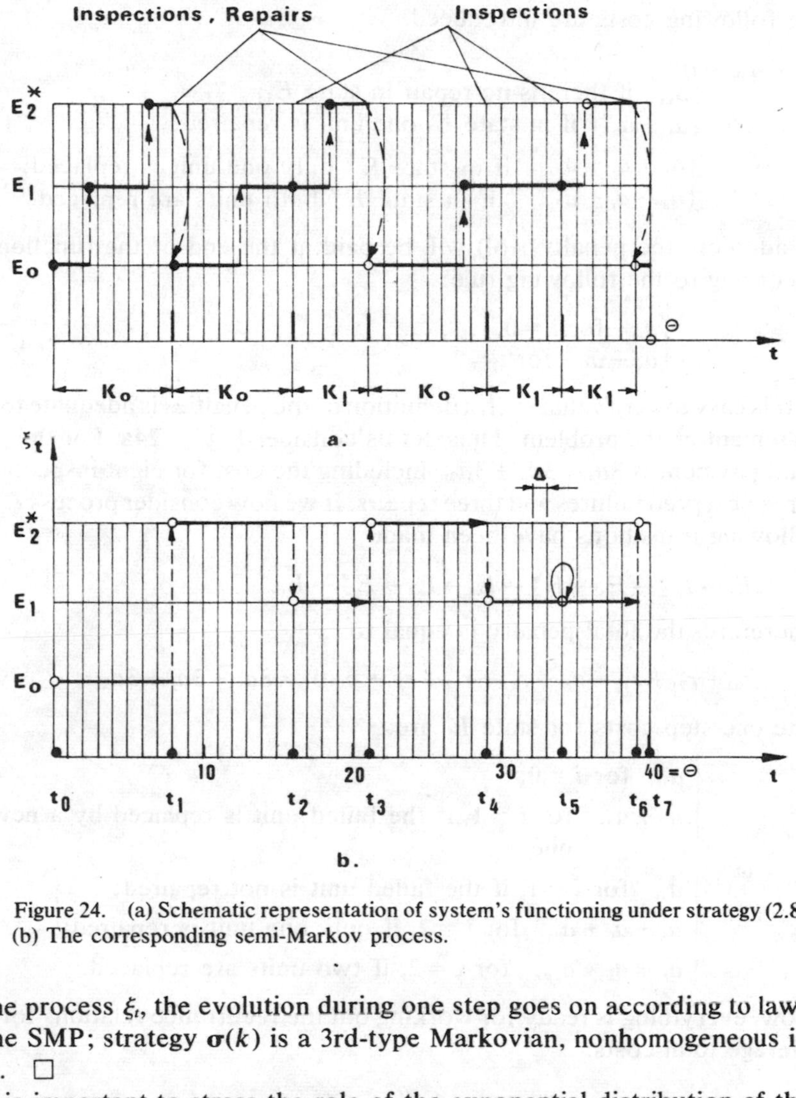

Figure 24. (a) Schematic representation of system's functioning under strategy (2.8). (b) The corresponding semi-Markov process.

In the process ξ_t, the evolution during one step goes on according to laws of the SMP; strategy $\sigma(k)$ is a 3rd-type Markovian, nonhomogeneous in time. □

It is important to stress the role of the exponential distribution of the time-to-failure of the unit. In the presence of the exponential distribution, the behaviour of the process ξ_t for $t > \mathbf{t}_n$ depends on the value $\mathbf{t}_n$, the state observed at moment $\mathbf{t}_n$ and does not depend on the behaviour before the time $\mathbf{t}_n$. This important circumstance is exploited for the evaluation of the recurrence relations for the costs in the process ξ_t.

Now, let us define the costs in the process ξ_t. For the transition $E_i \to E_j$

the following costs are introduced:

$$r_{0j} = a_s$$
$$r_{1j} = \begin{cases} a_s, & \text{if there is no repair in state } E_1; \\ a_s + a_p, & \text{if in state } E_1 \text{ one unit is repaired;} \end{cases} \tag{2.9}$$
$$r_{2j} = \begin{cases} a_s + a_j + a_p, & \text{if in state } E_2^* \text{ only one unit is replaced;} \\ a_s + a_j + a_{2p}, & \text{if in state } E_2^* \text{ both units are replaced.} \end{cases}$$

In addition, the penalty $v_i(0)$ will be paid at the end of the functioning according to the following rule:

$$v_i(0) = \begin{cases} a_s, & \text{for } i = 0, 1, \\ a_s + a_f, & \text{for } i = 2. \end{cases} \tag{2.10}$$

It is easy to verify that such a definition of the penalties is adequate to the statement of the problem. Thus, let us consider Figure 24a. For this, the total payment is $8a_s + 3a_p + 3a_{2p}$ including the cost for eight inspections, three observed failures and three repairs. If we now consider process ξ_t, the following transitions have been made

$$E_0 \to E_2^* \to E_1 \to E_2^* \to E_1 \to E_1 \to E_2^* \to E_0.$$

Therefore, the total penalty is equal to

$$r_{02} + r_{21} + r_{12} + r_{21} + r_{11} + r_{12} + r_{20} + v_0(0) = 8a_s + 3a_f + 3a_{2p}.$$

The one-step costs for state E_i are:

$$\omega_i = \begin{cases} a_s, & \text{for } i = 0; \\ a_s + a_p, & \text{for } i = 1, \text{ if the failed unit is replaced by a new one;} \\ a_s, & \text{for } i = 1, \text{ if the failed unit is not repaired;} \\ a_s + a_f + a_p, & \text{for } i = 2, \text{ if only one unit is repaired;} \\ a_s + a_f + a_{2p}, & \text{for } i = 2, \text{ if two units are replaced.} \end{cases}$$

Now, everything is ready for working out the recurrence relations for the average total costs.

2.3. Recurrence Relations for Costs in the Process ξ_t

Let $v_i(m)$ denote the minimal average cost which corresponds to the optimal strategy governing the process ξ_t, if the counting of costs starts at time $N - m$ and ends at the moment $\theta = N$ and the observed state at moment $N-m$ is E_i ($\xi_{N-m} = E_i$).† Assume that the values $v_i(1), \ldots, v_i(m-1)$, $i = 0, 1, 2$, are known. We shall demonstrate how the value $v_i(m)$ could be expressed.

First consider the case $m < K_1$ (it is supposed that $K_1 \leqslant K_0$). The total expected cost over the interval $(N-m, N)$ is equal to the sum of one-step cost for state E_i observed at instant $N-m$ plus the future expected cost. Besides that, the possibility to select one of many CAs available in state E_i must be taken into account. From here, according to the principle of optimality (see Section 3.4 of Chapter 1), we obtain the following relations:

$$
\begin{aligned}
v_0(m) &= a_s + \sum_{j=0}^{2} P_{0j}(m) v_j(0),\\
v_1(m) &= \min\Big\{a_s + \sum_{j=1}^{2} P_{1j}(m) v_j(0);\\
&\qquad a_s + a_p + \sum_{j=0}^{2} P_{0j}(m) v_j(0)\Big\}, \qquad (2.11)\\
v_2(m) &= \min\Big\{a_s + a_f + a_p + \sum_{j=1}^{2} P_{1j}(m) v_j(0);\\
&\qquad a_s + a_f + a_{2p} + \sum_{j=0}^{2} P_{0j}(m) v_j(0)\Big\}.
\end{aligned}
$$

By means of (2.11) it is possible to find in a recursive manner the values $v_i(m)$ for $m \leqslant K_1$. Now let m satisfy the inequality $K_1 < m \leqslant K_0$. Similar reasoning gives the following expressions:

$$
\begin{aligned}
v_0(m) &= a_s + \sum_{j=0}^{m} P_{0j}(m) v_j(0),\\
v_1(m) &= \min\Big\{a_s + \sum_{j=1}^{2} P_{1j}(K_1) v_j(m-K_1);\\
&\qquad a_s + a_p + \sum_{j=0}^{2} P_{0j}(m) v_j(0)\Big\}, \qquad (2.12)\\
v_2(m) &= \min\Big\{a_s + a_f + a_p + \sum_{j=1}^{2} P_{1j}(K_1) v_j(m-K_1);\\
&\qquad a_s + a_f + a_{2p} + \sum_{j=0}^{2} P_{0j}(m) v_j(0)\Big\}.
\end{aligned}
$$

†In the notation for costs only the starting state E_i and the remaining time m appear. This is correct because the behaviour of the observed process ξ_t depends only on these two parameters and does not depend on the "past" of the ξ_t, before time $t = N - m$.

It remains only to write the formulas for the case $m > K_0$:

$$
\begin{aligned}
v_0(m) &= a_s + \sum_{j=0}^{2} P_{0j}(K_0)v_j(m-K_0),\\
v_1(m) &= \min\Big\{a_s + \sum_{j=1}^{2} P_{1j}(K_1)v_j(m-K_1);\\
&\qquad a_s + a_p + \sum_{j=0}^{2} P_{0j}(K_0)v_j(m-K_0)\Big\}, \qquad (2.13)\\
v_2(m) &= \min\Big\{a_s + a_f + a_p + \sum_{j=1}^{2} P_{1j}(K_1)v_j(m-K_1);\\
&\qquad a_s + a_f + a_{2p} + \sum_{j=0}^{2} P_{0j}(K_0)v_j(m-K_0)\Big\}.
\end{aligned}
$$

It must be remarked that the "technology" of obtaining these recurrence relations was described in a general form in Chapter 1, Section 3.5.

2.4. Example

We shall now demonstrate how to use formulas (2.11)–(2.13) for finding the optimal strategy. The initial data are the following: the standby corresponds to the scheme of Figure 23a; $\lambda = 0.2$. The penalties $a_f = a_p = 0.5$; $a_{2p} = 2a_p = 1$; $a_s = 0.1$; the time period $N = 10$, $\gamma = 0.250$.

First the preparatory work must be done. In this particular case there is a simple relation between $v_2(m)$ and $v_1(m)$:

$$v_2(m) = a_f + a_p + v_1(m).$$

After substituting the value $v_2(m)$ into formulas for $v_1(m)$ and $v_0(m)$ we have

$$
\begin{aligned}
v_0(m) &= a_s + (a_f + a_p)P_{02}(x_0) + P_{00}(x_0)v_0(m-x_0)\\
&\quad + [1 - P_{00}(x_0)]v_1(m-x_0),\\
v_1(m) &= \min\{a_s + (a_f + a_p)P_{12}(x_1) + v_1(m-x_1);\ a_p + v_0(m)\},
\end{aligned} \qquad (2.14)
$$

where

$$x_1 = \min(m; K_1), \qquad x_0 = \min(m; K_0).$$

The well-known formulas for the "hot" standby are:

$$
\begin{aligned}
P_{00}(t) &= e^{-2\lambda t}, & P_{11}(t) &= e^{-\lambda t},\\
P_{01}(t) &= 2\,e^{-\lambda t}(1 - e^{-\lambda t}), & P_{12}(t) &= 1 - e^{-\lambda t}.\\
P_{02}(t) &= (1 - e^{-\lambda t})^2.
\end{aligned} \qquad (2.15)
$$

By means of (2.15) the following table is composed:

k	$P_{12}(k)$	$P_{02}(k)$	$P_{00}(k)$	$1-P_{00}(k)$
1	0.181	0.033	0.670	0.330
2	0.330	0.109	0.449	0.551
3	0.451	0.203	0.301	0.699
4	0.551	0.304	0.202	0.798

From this table we see that we have to take $K_0 = 3$; $K_1 = 1$. For these values of inspection time, the probability of a failure does not exceed the value 0.203. According to the initial data, $v_0(0) = v_1(0) = 0.1$; $v_2(0) = a_s + a_f = 0.6$. For $m = 1$, (2.14) have the form:

$$v_0(1) = a_s + (a_f + a_p)P_{02}(1) + P_{00}(1)v_0(0) + [1 - P_{00}(1)]v_1(0),$$

$$v_1(1) = \min\{a_s + (a_f + a_p)P_{12}(1) + v_1(0);\ a_p + v_0(1)\}.$$

Substituting the values of P_{ij}, a_s, a_j, a_p, we get:

$$v_0(1) = 0.1 + 0.033 + 0.1 = 0.233,$$

$$v_1(1) = \min\{0.1 + 0.181 + 0.1 = 0.381;\ 0.5 + 0.233 = 0.733\} = 0.381.$$

From the formula for $v_1(1)$ it follows that at the moment $N - m = 10 - 1 = 9$, the best policy for state E_1 is to do nothing, that is, $\sigma_1(N-1) \equiv \sigma_1(9) = \{E_1, 1\}$. For $m = 2$ we have:

$$v_0(2) = a_s + (a_f + a_p)P_{02}(2) + P_{00}(2)v_0(0) + [1 - P_{00}(2)]v_1(0),$$

$$v_1(2) = \min\{a_s + (a_f + a_p)P_{12}(1) + v_1(1);\ a_p + v_0(2)\}.$$

Substituting the values of P_{ij}, a_s, a_f, a_p and $v_1(1)$, $v_0(0)$ we obtain:

$$v_0(2) = 0.1 + 0.109 + 0.449 \cdot 0.1 + 0.551 \cdot 0.1 = 0.309,$$

$$v_1(2) = \min\{0.1 + 0.181 + 0{:}381 = 0.662;\ 0.5 + 0.309 = 0.809\} = 0.662.$$

Obviously, $\sigma_1(N-2) \equiv \sigma_1(8) = \{E_1, 1\}$. For $m = 3$ (2.14) have the form:

$$v_0(3) = a_s + (a_f + a_p)P_{02}(3) + P_{00}(3)v_1(0) + (1 - P_{00}(3))v_1(0),$$

$$v_1(3) = \min\{a_s + (a_f + a_p)P_{12}(1) + v_1(2);\ a_p + v_0(3)\}.$$

Substituting numbers, we obtain

$$v_0(3) = 0.403;\quad v_1(3) = \min(0.943;\ 0.903) = 0.903.$$

Now the situation has changed and $\sigma_1(N-3) = \sigma_1(7) = (E_0, 3)$. For $m \geqslant 4$

formulas (2.14) have the form:

$$v_0(m) = a_s + (a_f + a_p)P_{02}(3) + P_{00}(3)v_0(m-3) + [1 - P_{00}(3)]v_1(m-3)$$
$$= 0.303 + 0.301\,v_0(m-3) + 0.699\,v_1(m-3);$$
$$v_1(m) = \min\{a_s + (a_f + a_p)P_{12}(1) + v_1(m-1);\ a_p + v_0(m)\}$$
$$= \min\{0.281 + v_1(m-1);\ 0.5 + v_0(m)\}.$$

Using these expressions we obtain in a recursive manner:

$v_0(4) = 0.639$; $v_1(4) = 1.139$; $\sigma_1(6) = (E_0, 3)$; $v_0(5) = 0.859$; $v_1(5) =$
$v_0(7) = 1.291$; $v_1(7) = 1.791$; $\sigma_1(3) = (E_0, 3)$; $v_0(8) = 1.512$; $v_1(8) =$
$v_0(7) = 1.291$; $v_1(7) = 1.791$; $\alpha_1(3) = (E_0, 3)$; $v_0(8) = 1.512$; $v_1(8) =$
2.012; $\sigma_1(2) = (E_0, 3)$; $v_0(9) = 1.708$; $v_1(9) = 2.208$; $\sigma_1(1) = (E_0, 3)$;
$v_0(10) = 1.944$; $v_1(10) = 2.444$; $\sigma_1(0) = (E_0, 3)$.

From the relation $v_2(m) = a_f + a_p + v_1(m)$ it follows that the CA $\sigma_2(k)$ is an exact copy of the CA $\sigma_1(k)$.

Now, let us assume that the state at time $t = 0$ is E_1. Optimal strategy prescribes the following behaviour. At time $t = 0$, the repair $E_1 \to E_0$ has to be performed. The next inspection should be planned at moment $t = 3$. If during the inspection state E_2 or E_1 are discovered, the repair $E_i \to E_0$ must be done and the time of the next inspection would be $t = 6$. During this inspection, the actions prescribed by the optimal strategy remain the same. The third inspection has to be planned at time $t = 9$. In the case of failure, only one unit has to be repaired. If the state is E_1, nothing has to be done.

Concluding this section, we want to stress that the most important point in our consideration is the exact definition of the controlled process and its stochastic description. If the random process describing the behaviour of the system when it works "without outside interferences" is a Markovian one, then the observed process ξ_t preserves all the properties needed for writing the recurrence relations for the average costs. These properties, briefly speaking, guarantee that the controlled process ξ_t is a SMP. It is easy to see that all the formal construction we have done remains valid also for the system having an arbitrary number of units in the standby. It is also easy to introduce changes in the cost structure. Thus, one could consider the case of discounting which is often used when the reduction of all coming expenditures to the present moment has to be made. Formally speaking, it means that the cost paid at the moment $t = k$ must be multiplied by the discount factor α^k, $0 < \alpha < 1$. The book of

Jardine[49] is recommended to the reader for the detailed description of this discounting technique.

3. Determination of the Inspection Times for a Partially Repaired System

3.1. Statement of the Problem

A system having states E_0, E_1, E_2, E_3 and E_4^* is considered. The state E_4^* is an absorbing one and corresponds to the total failure of the system. The evolution of the system is described by a time-homogeneous Markov process $\eta(t)$. For the sake of simplicity it is assumed that the transitions in that process are done only from the "lower" states to the "upper" ones. The process $\eta(t)$ is defined by means of the following transition rates:

$$\underset{\Delta t\to 0}{P}\{\eta(t+\Delta t)=E_{i+1}|\eta(t)=E_i\}=\lambda_i\Delta t+o(\Delta t),\ i=0,1,2;$$

$$\underset{\Delta t\to 0}{P}\{\eta(t+\Delta t)+E_4^*|\eta(t)=E_j\}=\mu_j\Delta t+o(\Delta t),\ j=0,1,2,3;$$

$$\underset{\Delta t\to 0}{P}\{\eta(t+\Delta t)=E_i\,|\,\eta(t)=E_i\}$$

$$=\begin{cases}1-(\lambda_i+\mu_i)\Delta t+o(\Delta t), & i=0,1,2;\\ 1-\mu_3\Delta t+o(\Delta t), & i=3.\end{cases}\tag{3.1}$$

The scheme of transitions is given in Figure 25. The system cannot move from the failure state E_4^*, which is the "trapping state". Formally, for all $\tau>0$

$$P\{\eta(t+\tau)=E_4^*|\eta(t)=E_4^*\}\equiv 1.$$

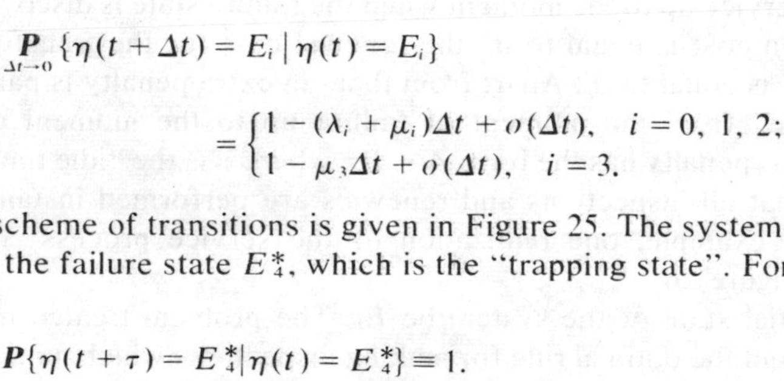

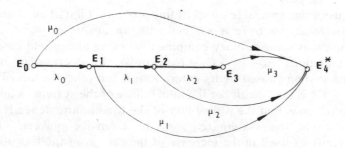

Figure 25. Transition diagram for the process $\eta(t)$.

Note that relation (3.1) describes the evolution of the process only locally, over a small time interval $(t, t + \Delta t)$. However, if the initial state E_r at time $t = 0$ is known, one can obtain the whole probabilistic description for the process $\eta(t)$. For example, by means of a system of differential equations, we can derive the values of transition probabilities $P_{ij}(t)$, $i = 0, \ldots, 4$. For the evaluation of these equations, it is enough to know the rates (3.1). Usually, in solving applied problems, the information concerning the transition rates over a small time interval $(t, t + \Delta t)$ is the starting point for finding all characteristics of the considered Markov process. As an example, one could mention the birth-and-death process[32, Chapter 17] widely used in both queueing and reliability theories.

For the system under consideration, the following rule of service is accepted. From time to time the system is inspected; during the inspection, the actual state of the system is ascertained. If the state E_4^* appears, the work of the system is completed. If the "critical" state E_3 is detected, a partial renewal is done which means that the system is transferred into state E_2. If states E_0, E_1 or E_2 are detected, nothing has to be done and the next inspection time is fixed respectively, after periods T_0, T_1 or T_2. The system is in service up to the moment when the failure state is discovered. The inspection cost is equal to a_s, the renewal cost for the transferring from E_3 to E_2 is equal to a_p. Apart from that, an extra penalty is paid for the "idle time" from the moment of failure up to the moment of its discovery. This penalty has the form $A + B\tau$, where τ is the "idle time". It is assumed that all inspections and renewals are performed instantaneously. As an example, one realization of the service process is represented in Figure 26.

Let the initial state of the system be E_0. The problem treated in this section is to find the optimal rule for making inspections which minimizes the total expected cost.

Let us discuss some special features of this problem. First of all, it must be pointed out that we have a system with an absorbing state. An excellent example is some military equipment working under field conditions. The transitions $E_0 \to E_1 \to E_2 \to E_3$ express the decrease of the abilities of the system caused by its deterioration and other factors. The transition $E_3 \to E_4^*$ is the "death" or the total failure of the system. A quite realistic supposition is that the probability of the total failure depends on the actual state of the system (e.g. on its defensive power). This peculiarity manifests itself in the increase of the rate μ_i of the transition $E_i \to E_4^*$. In the problem under consideration, we restrict ourselves only

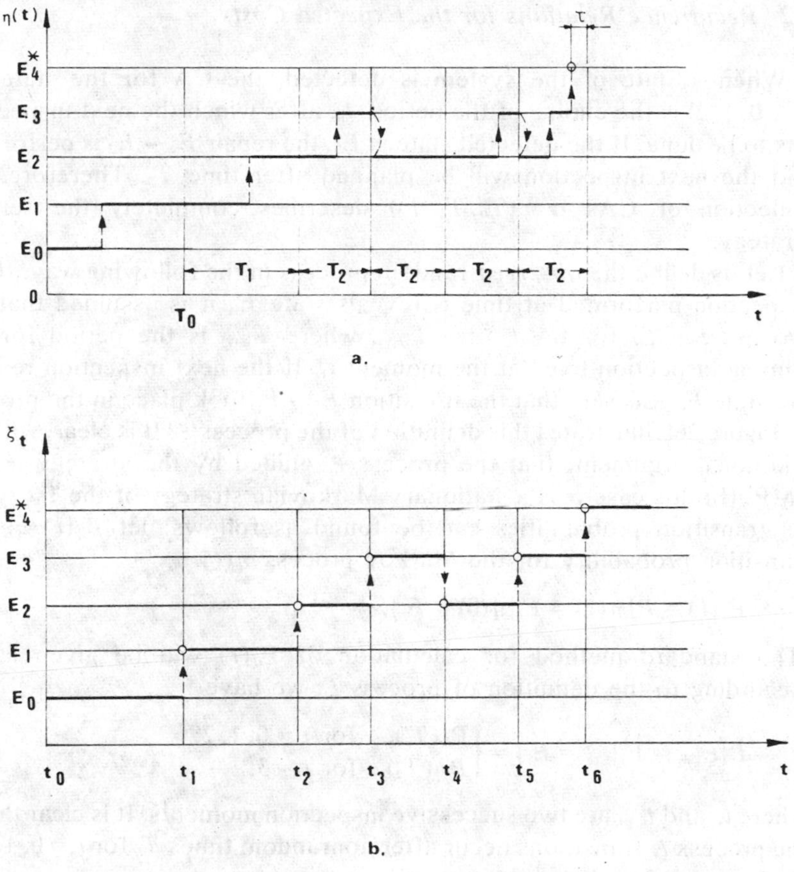

Figure 26. (a) A record of system's functioning. (b) The observed process ξ_t.

to minimizing the total cost which depends on the number of inspections, repairs and the length of the "idle" period. Such a statement of the problem makes sense for very responsible systems which have relatively high penalties for failure, idle time and repairs.

The formal aspect of the treated problem has two following important features: the service process is not restricted by a prescribed time interval, and the time for the next inspection can be chosen in such a way that it depends on the last detected state of the system. Similar problems have been considered in [6] and [35].

3.2. Recurrence Relations for the Expected Costs

When a state of the system is detected, the CA for the states E_i $(i = 0, 1, 2)$ is the choice of the period T_i, after which the next inspection has to be done. If the detected state is E_3, the repair $E_3 \to E_2$ is performed and the next inspection will be planned after time T_2. Therefore, the collection of CAs $\boldsymbol{\sigma} = (T_0, T_1, T_2)$ describes completely the service strategy.

Let us define the *observed* random process in the following way. If the inspection performed at time $\mathbf{t}_l$ reveals state E_i, it is assumed that the process $\xi_t = E_i$ for $\mathbf{t}_l \leq t < \mathbf{t}_l + T_{\varphi(i)}$, where $T_{\varphi(i)}$ is the period for the coming inspection fixed at the moment $\mathbf{t}_l$. If the next inspection reveals the state E_j, it is said that the transition $E_i \to E_j$ took place in the process ξ_t. Figure 26b illustrates this definition of the process ξ_t. It is clear, without additional argument, that the process ξ_t guided by the strategy $\boldsymbol{\sigma}$ is a SMP. (In this case $\boldsymbol{\sigma}$ is a stationary Markovian strategy of the 1st type.) Its transition probabilities can be found as follows: let $P_{ij}(t)$ be the transition probability for the Markov process $\eta(t)$:

$$P_{ij}(t) = \boldsymbol{P}\{\eta(t) = E_j \,|\, \eta(0) = E_i\}. \tag{3.2}$$

(The standard method for calculation of $P_{ij}(t)$ will be given later.) According to the definition of process ξ_t, we have

$$\boldsymbol{P}\{\xi_{t_{k+1}} = E_j \,|\, \xi_{t_k} = E_i\} = \begin{cases} P_{ij}(T_i), & \text{for } i = 0, 1, 2; \\ P_{2j}(T_2), & \text{for } i = 3, \end{cases} \tag{3.3}$$

where t_k and t_{k+1} are two successive inspection moments. It is clear that in the process ξ_t transitions occur after nonrandom times T_i for $i = 0, 1$, and T_2 for $i = 2, 3$.

Now let us determine the costs for process ξ_t. Assume that the transition $E_i \to E_4^*$ has taken place $(i = 0, 1, 2)$ and calculate the costs r_{i4} corresponding to that transition. The probability that the system will fail in the interval $(t, t + \Delta t)$, calculated under the condition that a failure occurred in the interval $(0, T_i)$, is equal to

$$\frac{P_{i4}(t + \Delta t) - P_{i4}(t)}{P_{i4}(T_i)} = \frac{P'_{i4}(t)\Delta t}{P_{i4}(T_i)} + o(\Delta t), \qquad \Delta t \to 0. \tag{3.4}$$

According to the statement of the problem, the penalty $A + B(T_i - t)$ corresponding to the "idle time" $(T_i - t)$ has to be counted. In addition, the cost a_s must be paid for the inspection itself. Therefore, the expected cost for the transition $E_i \to E_4^*$ is

$$r_{i4} = a_s + \int_0^{T_i} \frac{A + B(T_i - t)}{P_{i4}(T_i)} P'_{i4}(t)\,\mathrm{d}t. \tag{3.5}$$

The payment for the transitions from states E_0, E_1 and E_2 into E_j $(j \neq 4)$ consists only of the inspection cost. Therefore,

$$r_{0j} = r_{1j} = r_{2j} = a_s, \qquad j \neq 4. \tag{3.6}$$

According to (3.3), (3.5) and (3.6) we get the one-step costs $\omega_i(T_i)$ for the states E_i, $i = 0, 1, 2$:

$$\omega_i(T_i) = \sum_{j=i}^{4} r_{ij} P_{ij}(T_i) = a_s + \int_0^{T_i} [A + B(T_i - t)] P'_{i4}(t)\,\mathrm{d}t. \tag{3.7}$$

It remains now to calculate the one-step cost for state E_3. According to the given description, for E_3 the additional cost a_p is paid:

$$\omega_3(T_2) = a_p + \omega_2(T_2). \tag{3.8}$$

Formula (3.7) can be simplified and rewritten in the form:

$$\omega_i(T_i) = a_s + AP_{i4}(T_i) + B \int_0^{T_i} P_{i4}(t)\,\mathrm{d}t, \qquad i = 0, 1, 2. \tag{3.9}$$

Up to now we have not taken into account the penalty paid for the last inspection when the failure of the system is detected. However, this does not change anything in essence because the service process sooner or later will be finished and the addition of a constant value does not change the situation.

Let state E_i be the initial one in process ξ_t. It will be assumed that there exists the optimal stationary strategy $\boldsymbol{\sigma} = (T_0^*, T_1^*, T_2^*)$ which provides the minimal value of the expected total cost. Let V_i denote the minimal cost for the process ξ_t starting in state E_i at time $t = 0$. Since for every strategy $\boldsymbol{\sigma}$ the process ξ_t has a finite duration with probability 1, the values V_i are finite. To avoid some difficulties of a formal nature, it will be assumed that T_i can have only a finite set of different values. Then the total number of stationary strategies is finite and it is obvious that an optimal strategy does exist. In our problem, adding of the costs takes place up to the moment of discovering the failure. This situation was described in a general manner in Section 3 of Chapter 1 ("The process with an absorbing state".) Repeating the reasoning given there, we obtain the following system of recurrence relations for V_i:

$$V_i = \min_{0<T_i}\left[\omega_i(T_i) + \sum_{j=i}^{3} P_{ij}(T_i)V_j\right], \qquad i = 0, 1, 2;$$
$$V_3 = \min_{0<T_2}\left[\omega_3(T_2) + \sum_{j=2}^{3} P_{2j}(T_2)V_2\right] = a_p + V_2. \tag{3.10}$$

It is clear that we can exclude the last relation by substituting the value $V_3 = V_2 + a_p$ in the first three relations. After that we have the following system:

$$V_0 = \min_{0<T_0}\left[\omega_0(T_0) + a_pP_{03}(T_0) + P_{03}(T_0)V_2 + \sum_{j=0}^{2} P_{0j}(T_0)V_j\right],$$
$$V_1 = \min_{0<T_1}\left[\omega_1(T_1) + a_pP_{13}(T_1) + P_{13}(T_1)V_2 + \sum_{j=1}^{2} P_{1j}(T_1)V_j\right], \tag{3.11}$$
$$V_2 = \min_{0<T_2}[\omega_2(T_2) + (a_p + V_2)P_{23}(T_2) + P_{22}(T_2)V_2].$$

3.3. *Computation of Transition Probabilities $P_{ij}(t)$ in the Process $\eta(t)$*

Let $\eta(t)$ be a time-homogeneous Markov process with a finite number of states $E_0, E_1, \ldots, E_n$. It is assumed that the transition rates λ_{kj} are known:

$$\mathbf{P}\{\eta(t+\Delta t) = E_j \mid \eta(t) = E_k\} = \lambda_{kj}\Delta t + o(\Delta t), \qquad j \neq k,$$
$$\underset{\Delta t\to 0}{\mathbf{P}}\{\eta(t+\Delta t) = E_j \mid \eta(t) = E_j\} = 1 - \sum_{j\neq k} \lambda_{kj}\Delta t + o(\Delta t). \tag{3.12}$$

Let the initial state at $t = 0$ be E_i: $\mathbf{P}\{\eta(0) = E_i\} = 1$. We shall describe a standard procedure for finding values

$$P_{ij}(t) = \mathbf{P}\{\eta(t) = E_j \mid \eta(0) = E_i\}.$$

Let us introduce the following matrix $\mathbf{B}$

$$\mathbf{B} = \begin{Vmatrix} \lambda_0^* & \lambda_{01} & \lambda_{02} & \cdots & \lambda_{0n} \\ \lambda_{10} & \lambda_1^* & \lambda_{12} & \cdots & \lambda_{1n} \\ \cdot & \cdot & \cdot & \cdots & \cdot \\ \cdot & \cdot & \cdot & \cdots & \cdot \\ \cdot & \cdot & \cdot & \cdots & \cdot \\ \lambda_{n0} & \lambda_{n1} & \lambda_{n2} & \cdots & \lambda_n^* \end{Vmatrix}. \tag{3.13}$$

Here

$$\lambda_k^* = -\sum_{j \neq k} \lambda_{kj}, \qquad k = 0, \ldots, n.$$

Let $\mathbf{P}_i'(t)$ and $\mathbf{P}_i(t)$ be columns with elements $P'_{i0}(t), \ldots, P'_{in}(t)$ and $P_{i0}(t), \ldots, P_{in}(t)$ respectively. The transition probabilities $P_{ij}(t)$ satisfy the following system of differential equations [19, Chapter 6]:

$$\mathbf{P}_i'(t) = \mathbf{B}^T \mathbf{P}_i(t) \tag{3.14}$$

with the initial condition

$$P_{ik}(0) = \begin{cases} 0, & k \neq i, \\ 1, & k = i. \end{cases}$$

$\mathbf{B}^T$ is the transposed matrix $\mathbf{B}$.

The method which usually simplifies the solution of the system (3.14) consists of applying the Laplace transform, which leads to a system of linear algebraic equations. Let us denote

$$\pi_j(s) = \int_0^\infty e^{-st} P_{ij}(t)\, dt. \tag{3.15}$$

Applying the transformation (3.15) to each equation of (3.14) and using the formula

$$\int_0^\infty e^{-st} P'_{ij}(t)\, dt = s\pi_j(s) - P_{ij}(0), \tag{3.16}$$

we obtain the system

$$s\Pi(s) - \mathbf{P}_i(0) = \mathbf{B}^T \Pi(s),$$

where $\Pi(s)$ is the column consisting of elements $\pi_j(s)$, $j = 0, \ldots, n$. From this follows the system of equations

$$(\mathbf{I}s - \mathbf{B}^T)\Pi(s) = \mathbf{P}_i(0), \tag{3.17}$$

where $\mathbf{I}$ is the identity matrix.

Now this system must be solved in respect to $\pi_j(s)$ and the functions $P_{ij}(t)$ must be found with the help of Laplace transform tables.

We note that in our situation one or many states will be absorbing. If E_k is an absorbing state, then $\lambda_{kj} = 0$ for $j \neq k$. The row in matrix $\mathbf{B}$ which corresponds to the absorbing state E_k has to consist only of zero elements.

We shall illustrate how the general method works with an example of the system represented in Figure 25. The initial state at $t = 0$ is E_0. We note that state E_4 is an absorbing one. The matrix **B** has the form

$$\mathbf{B} = \begin{Vmatrix} -(\lambda_0+\mu_0) & \lambda_0 & 0 & 0 & \mu_0 \\ 0 & -(\lambda_1+\mu_1) & \lambda_1 & 0 & \mu_1 \\ 0 & 0 & -(\lambda_2+\mu_2) & \lambda_2 & \mu_2 \\ 0 & 0 & 0 & -\mu_3 & \mu_3 \\ 0 & 0 & 0 & 0 & 0 \end{Vmatrix}. \tag{3.18}$$

The system (3.17) is:

$$\begin{aligned} &(s+\lambda_0+\mu_0)\pi_0(s) = 1, \\ &-\lambda_0\pi_0(s) + (s+\lambda_1+\mu_1)\pi_1(s) = 0, \\ &-\lambda_1\pi_1(s) + (s+\lambda_2+\mu_2)\pi_2(s) = 0, \\ &-\lambda_2\pi_2(s) + (s+\mu_3)\pi_3(s) = 0, \\ &-\mu_0\pi_0(s) - \mu_1\pi_1(s) - \mu_2\pi_2(s) - \mu_3\pi_3(s) + s\pi_4(s) = 0. \end{aligned} \tag{3.19}$$

It is enough to find $\pi_j(s)$ for $j = 0, 1, 2, 3$. Solving equations starting with the first, we obtain

$$\begin{aligned} &\pi_0(s) = (s+\lambda_0+\mu_0)^{-1}, \qquad \pi_1(s) = \lambda_0 \prod_{i=0}^{1} (s+\lambda_i+\mu_i)^{-1}, \\ &\pi_2(s) = \lambda_0\lambda_1 \prod_{i=0}^{2} (s+\lambda_i+\mu_i)^{-1}, \\ &\pi_3(s) = \lambda_0\lambda_1\lambda_2 \prod_{i=0}^{2} (s+\lambda_i+\mu_i)^{-1}(s+\mu_3)^{-1}. \end{aligned} \tag{3.20}$$

It remains only to use the tables of inverse Laplace transform for finding the functions $P_{0j}(t)$ (see [30, Appendix]). It will be assumed that all values $(\lambda_i+\mu_i)$, μ_3 are different. In that case, formulas (35, 44 and 61) in the Appendix of [30] are applicable. Denoting $\mu_0+\lambda_0 = a_0$, $\mu_1+\lambda_1 = a_1$, $\mu_2+\lambda_2 = a_2$, $\mu_3 = a_3$, we obtain:

$$P_{00}(t) = e^{-a_0 t}; \qquad P_{01}(t) = \frac{\lambda_0}{a_1-a_0} e^{-a_0 t} + \frac{\lambda_0}{a_0-a_1} e^{-a_1 t},$$

$$P_{02}(t) = \frac{\lambda_0\lambda_1 e^{-a_0 t}}{(a_1-a_0)(a_2-a_0)} + \frac{\lambda_0\lambda_1 e^{-a_1 t}}{(a_2-a_1)(a_0-a_1)} + \frac{\lambda_0\lambda_1 e^{-a_2 t}}{(a_1-a_2)(a_0-a_2)},$$

$$\begin{aligned} P_{03}(t) = &\frac{\lambda_0\lambda_1\lambda_2 e^{-a_0 t}}{(a_3-a_0)(a_2-a_0)(a_1-a_0)} + \frac{\lambda_0\lambda_1\lambda_2 e^{-a_1 t}}{(a_3-a_1)(a_2-a_1)(a_0-a_1)} \\ &+ \frac{\lambda_0\lambda_1\lambda_2 e^{-a_2 t}}{(a_3-a_2)(a_1-a_2)(a_0-a_2)} + \frac{\lambda_0\lambda_1\lambda_2 e^{-a_3 t}}{(a_0-a_3)(a_1-a_3)(a_2-a_3)}. \end{aligned}$$

Transition probabilities $P_{1j}(t)$ and $P_{2j}(t)$ can be found in a similar manner. We omit the routine calculation and give the results:

$$P_{11}(t) = e^{-a_1 t},$$

$$P_{12}(t) = \frac{\lambda_1}{a_2 - a_1} e^{-a_1 t} + \frac{\lambda_1}{a_1 - a_2} e^{-a_2 t},$$

$$P_{13}(t) = \frac{\lambda_1 \lambda_2}{(a_3 - a_1)(a_2 - a_1)} e^{-a_1 t} + \frac{\lambda_1 \lambda_2}{(a_3 - a_2)(a_1 - a_2)} e^{-a_2 t} + \frac{\lambda_1 \lambda_2}{(a_1 - a_3)(a_2 - a_3)} e^{-a_3 t},$$

$$P_{22}(t) = e^{-a_2 t}, \qquad P_{23}(t) = \frac{\lambda_2}{a_3 - a_2} e^{-a_2 t} + \frac{\lambda_2}{a_2 - a_3} e^{-a_3 t}.$$

3.4. Example for the Computation of the Optimal Strategy

We give a numerical example which illustrates the method of finding the optimal inspection strategy. Let $\lambda_0 = \lambda_1 = \lambda_2 = 1$, $\mu_0 = 0$, $\mu_1 = 0.5$, $\mu_2 = 1.5$, $\mu_3 = 3$. Then the formulas for $P_{ij}(t)$ are:

$$P_{00}(t) = e^{-t}, \qquad P_{01}(t) = 2e^{-t} - 2e^{-1.5t},$$

$$P_{02}(t) = 1.333\, e^{-t} - 2e^{-1.5t} + 0.667\, e^{-2.5t};$$

$$P_{03}(t) = 0.667\, e^{-t} - 1.333\, e^{-1.5t} + 1.333\, e^{-2.5t} - 0.667\, e^{-3t},$$

$$P_{11}(t) = e^{-1.5t}, \qquad P_{12}(t) = e^{-1.5t} - e^{-2.5t},$$

$$P_{13}(t) = 0.667\, e^{-1.5t} - 2\, e^{-2.5t} + 1.333\, e^{-3t}, \qquad P_{22}(t) = e^{-2.5t},$$

$$P_{23}(t) = 2\, e^{-2.5t} - 2\, e^{-3t}. \; (a_0 = 1,\ a_1 = 1.5,\ a_2 = 2.5,\ a_3 = 3).$$

To calculate the costs we need the values $P_{i4}(t)$ for $i = 0, 1, 2$. They are:

$$P_{04}(t) = 1 - \sum_{j=0}^{3} P_{0j}(t) = 1 - 5\, e^{-t} + 5.333\, e^{-1.5t} - 2\, e^{-2.5t} + 0.667\, e^{-3t},$$

$$P_{14}(t) = 1 - \sum_{j=1}^{3} P_{1j}(t) = 1 - 2.667\, e^{-1.5t} + 3\, e^{-2.5t} - 1.333\, e^{-3t},$$

$$P_{24}(t) = 1 - \sum_{j=2}^{3} P_{2j}(t) = 1 - 3\, e^{-2.5t} + 2\, e^{-3t}.$$

Let $a_s = a_p = 1$, $A = 0$, $B = 1$, and calculate the one-step costs $\omega_i(T_i)$

according to (3.9):

$$\omega_0(T_0) = T_0 - 1.022 + 5.000\,e^{-T_0} - 3.556\,e^{-1.5T_0} + 0.8\,e^{-2.5T_0} - 0.222\,e^{-3T_0},$$

$$\omega_1(T_1) = T_1 - 0.022 + 1.778\,e^{-1.5T_1} - 1.200\,e^{-2.5T_1} + 0.444\,e^{-3T_1},$$

$$\omega_2(T_2) = T_2 + 1.200\,e^{-2.5T_2} - 0.667\,e^{-3T_2} + 0.467.$$

Now let us move to the direct computation of the optimal strategy. We shall act according to the algorithm described in Section 3.6 of Chapter 1.

1st step. *Construction of the initial approximation for the sequence which converges to the solution from above.* Let us take an arbitrary strategy $\boldsymbol{\sigma}^{(0)} = (T_0^{(0)}, T_1^{(0)}, T_2^{(0)}) = (3, 2, 1)$ and find the expected costs which correspond to that strategy. For that purpose we solve the system (3.11) omitting the "min" on the right side and setting $T_0 = 3$, $T_1 = 2$, $T_2 = 1$. After simple transformation, (3.11) takes the form:

$$V_0[1 - P_{00}(T_0)] + V_1(-P_{01}(T_0)) + V_2(-P_{02}(T_0) - P_{03}(T_0)) = \omega_0(T_0) + P_{03}(T_0),$$

$$V_1(1 - P_{11}(T_1)) + V_2(-P_{12}(T_1) - P_{13}(T_1)) = \omega_1(T_1) + P_{13}(T_1),$$

$$V_2(1 - P_{22}(T_2) - P_{23}(T_2)) = \omega_2(T_2) + P_{23}(T_2).$$

After counting the coefficients, the system will have the following form:

$$\begin{aligned} V_0 \cdot 0.9502 + V_1(-0.0744) + V_2(-0.0636) &= 2.207\\ V_1 \cdot 0.9502 \qquad\qquad + V_2(-0.0661) &= 2.083\\ V_2 \cdot 0.8533 &= 1.597 \end{aligned}$$

Its solution is: $\hat{V}_0^{(0)} = 2.637$; $\hat{V}_1^{(0)} = 2.322$; $\hat{V}_2^{(0)} = 1.872$.

2nd step. *Construction of the sequence converging to the solution from above.* Let us substitute the values ω_i and P_{ij} in the system (3.11). After some transformations, the system becomes the following form which is convenient for further calculations:

$$\begin{aligned} \hat{V}_0^{(k+1)} = \min_{T_0} \{&-1.022 + T_0\\ &+ e^{-T_0}[5.667 + 1.000\,\hat{V}_0^{(k)} + 2.000\,\hat{V}_1^{(k)} + 2.000\,\hat{V}_2^{(k)}]\\ &+ e^{-1.5T_0}[-4.889 - 2.000\,\hat{V}_1^{(k)} - 3.333\,\hat{V}_2^{(k)}]\\ &+ e^{-2.5T_0}[2.133 + 2.000\,\hat{V}_2^{(k)}] + e^{-3T_0}[-0.889 - 0.667\,\hat{V}_2^{(k)}]\}, \end{aligned}$$

$$\begin{aligned} \hat{V}_1^{(k+1)} = \min_{T_1} \{&-0.022 + T_1 + e^{-1.5T_1}[2.445 + 1.000\,\hat{V}_1^{(k)} + 1.667\,\hat{V}_2^{(k)}]\\ &+ e^{-2.5T_1}[-3.200 - 3.000\,\hat{V}_2^{(k)}] + e^{-3T_1}[1.777 + 1.333\,\hat{V}_2^{(k)}]\}, \end{aligned}$$

$$\hat{V}_2^{(k+1)} = \min_{T_2} \{0.467 + T_2 + e^{-2.5T_2}[3.200 + 3.000\,\hat{V}_2^{(k)}] + e^{-3T_2}[-2.667 - 2.000\,\hat{V}_2^{(k)}]\}.$$

Substituting the vector $\hat{\mathbf{V}}^{(0)}$, we find the vector $\hat{\mathbf{V}}^{(1)}$. Searching for the minimum is done on the discrete time scale, $T = 0.01\ (0.01)\ 3.00$, with a computer used to perform the calculations. The vector $\tilde{\mathbf{V}}_1^{(1)}$ is equal to

$$\hat{\mathbf{V}}^{(1)} = (2.466;\ 2.141;\ 1.856),$$

and the corresponding values of T_i which minimize the right sides of the recurrence equations are $\hat{T}_0^{(1)} = 2.25$, $\hat{T}_1^{(1)} = 1.38$, $\hat{T}_2^{(1)} = 0.92$. Substituting the vector $\hat{\mathbf{V}}^{(1)}$ we calculate the vector $\hat{\mathbf{V}}^{(2)}$, and so on. The sequence of $\hat{\mathbf{V}}^{(k)}$ is the following:

$$\hat{\mathbf{V}}^{(2)} = (2.420;\ 2.117;\ 1.864),$$
$$\hat{\mathbf{V}}^{(3)} = (2.411;\ 2.114;\ 1.864),$$
$$\hat{\mathbf{V}}^{(4)} = (2.411;\ 2.113;\ 1.864),$$
$$\hat{\mathbf{V}}^{(5)} = (2.410;\ 2.113;\ 1.864).$$

The following values of $\hat{T}_i^{(5)}$ correspond to the vector $\hat{\mathbf{V}}^{(5)}$:

$$\hat{T}_0^{(5)} = 2.17, \qquad \hat{T}_1^{(5)} = 1.34, \qquad \hat{T}_2^{(5)} = 0.92.$$

3rd step. *Construction of the sequence* $\mathbf{V}^{(k)}$ *converging to the solution from below.* Let us substitute in the system (3.11) zero-vector instead of the vector $\hat{\mathbf{V}}^{(0)}$. Acting in a manner similar to the above iterative procedure, we obtain an increasing sequence of vectors $\mathbf{V}^{(k)}$:

$$\mathbf{V}^{(1)} = (1.000;\ 1.000;\ 1.000), \qquad \mathbf{V}^{(5)} = (2.403;\ 2.110;\ 1.863),$$
$$\mathbf{V}^{(2)} = (1.859;\ 1.761;\ 1.635), \qquad \mathbf{V}^{(6)} = (2.408;\ 2.112;\ 1.863),$$
$$\mathbf{V}^{(3)} = (2.262;\ 2.034;\ 1.829), \qquad \mathbf{V}^{(7)} = (2.409;\ 2.113;\ 1.864).$$
$$\mathbf{V}^{(4)} = (2.376;\ 2.097;\ 1.857),$$

The values of $T_i^{(7)}$, optimal for the vector $\mathbf{V}^{(7)}$, are:

$$T^{(7)} = (2.17,\ 1.34,\ 0.92).$$

These values coincide with $\hat{T}_i^{(5)}$. If it is assumed that two decimal places provide the needed precision, we obtain the following solution: the vector of minimal average costs is

$$\mathbf{V} = (2.41,\ 2.11,\ 1.86)$$

and the optimal strategy is

$$\boldsymbol{\sigma} = (2.17,\ 1.34,\ 0.92).$$

In this example it is sufficient to perform four iterations from above and below to achieve a solution which differs from the optimal one by less than 0.01.

3.5. Possible Generalizations

It would be worthwhile to consider, at this time, two possible generalizations of the problem which was presented above. From the practical point of view, it is necessary to take into account some additional costs which may be caused by the system's sojourn in "intermediate" states. For example, if a system is performing a specific job, its productivity might decrease when the system deteriorates – or formally – when it goes from state E_0 to higher states. This situation could be taken into account by introducing penalties (costs) depending on the time the system has spent in each state. A good example is a multichannelled commutational telephone system. "Deterioration" of this system is simply a failure of its channels, which leads to an increase in the number of lost telephone calls. If a certain cost is associated with every lost call, the additional payment would be proportional to the average number of lost calls. An example of this sort is discussed in [44]. Speaking formally, changes in the structure of costs will alter only the expression for ω_i. In Problem 4, Section 10 of this chapter we will discuss how to execute the needed calculations.

The second generalization deals with the consideration of a wider class of service strategies. Indeed, up to now we have made an agreement to do nothing in states E_0, E_1, E_2 and to do only a partial repair $E_3 \to E_2$ for state E_3. We might postulate that for every state (except E_0 and E_4) all kinds of repairs are available. Formally, this means that the system from state E_i can be, after an inspection, transferred into state E_j $(j < i)$, for an additional cost of c_{ij}, of course. We shall show how this change is reflected in the recurrence relations (3.10). Let us assume that in state E_2 we can do two kinds of repairs: $E_2 \to E_0$, $E_2 \to E_1$, or nothing. The cost for the repair $E_i \to E_j$ should be equal to c_{2i}, $i = 0, 1$. Then the recurrence relation for state E_2 will have the form:

$$V_2 = \min \left\{ \min_{T_2 > 0} \left[\omega_2(T_2) + \sum_{j=2}^{3} P_{2j}(T_2) V_j \right];\right.$$

$$\min_{T_2>0}\left[c_{21}+\omega_1(T_2)+\sum_{j=1}^{3}P_{1j}(T_2)V_j\right];$$
$$\left.\min_{T_2>0}\left[c_{20}+\omega_0(T_2)+\sum_{j=0}^{3}P_{0j}(T_2)V_j\right]\right\}. \quad (3.10^*)$$

The pair $(E_i \to E_{\varphi(i)}; T_{\varphi(i)})$ appears now as a CA for state E_2. If in formula (3.10*) the optimal value is $T_2 = T_2^*$, and the minimum was found, for example, in the second row, this means that the optimal CA for state E_2 is the repair "$E_2 \to E_1$" and the appointment for the coming inspection after time period T_2^*. The changes in the iterative procedure are obvious and they only slightly alter the process of finding the optimal strategy.

4. Optimal Use of Standby Elements

4.1. Statement of the Problem

Let us consider a system of one main and $(r-1)$ standby elements. It has an important peculiarity: the number of active standby elements may change during the course of its operation. More precisely, at the moment $t=0$ it is possible to engage an arbitrary number j, $j=1,\ldots,r$, of elements (the standby redundancy is assumed to be "hot", see Figure 27). Those elements which are not switched into operation are not subjected to failures. The rate of failures is assumed to be constant for every working element, $\lambda = \text{Const}$. Some time moments $t_1, t_2, \ldots, t_m, t_{m+1}$ are

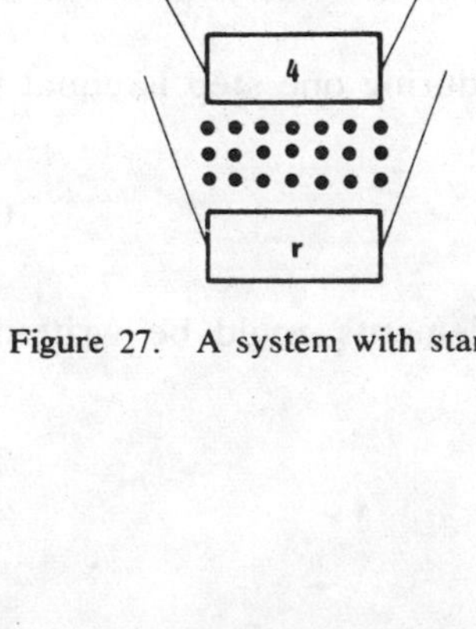

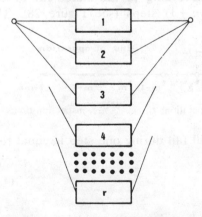

Figure 27. A system with standby redundancy.

appointed beforehand for an inspection. If the inspection (done at the time t_k) reveals that all elements which had started to work at time t_{k-1} have failed, then this is considered as a total failure of the system and the system stops functioning. Otherwise, the total number r_1 of nonfailed elements is counted and a number l, $1 \leqslant l \leqslant r_1$, of them are switched on for a one time period, that is, for (t_k, t_{k+1}). It will be assumed also that the time for inspections and switching on are negligible quantities. Eventually, the failed elements cannot be renewed.

How can we determine the number of elements which, when put into operation at every inspection, minimize the probability of total failure occurring at interval $(0, t_{m+1})$?

It is clear that at moment $t = t_m$ all nonfailed elements must be operating. However, if "in the future" there will be many inspections, it is not clear whether or not it is expedient to switch on all acting elements. If, for example, the maximal possible number of elements is switched on for one-time period (t_s, t_{s+1}), we have a minimal probability of a total failure during this interval. At the same time, that would speed up the "deterioration" of the system and therefore could lead to the increase of probability of the total failure during the remaining interval (t_{s+1}, t_{m+1}).

4.2. Recurrence Relations for the Probability of Total Failure

It is assumed that inspections occur at equally spaced moments, $t_{i+1} - t_i = \Delta$, $i = 0, \ldots, m$. The interval between two adjacent inspections will be called a step. Therefore, according to the statement of the problem, the system must work $(m+1)$ steps (see Figure 28). The

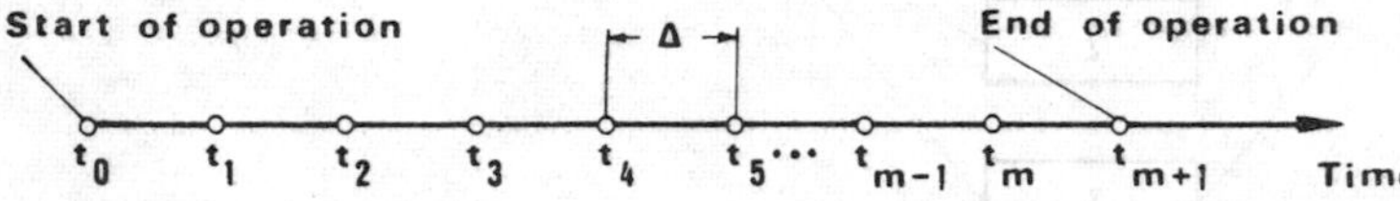

Figure 28. Scheme of $(m+1)$ step operation; $t_1, t_2, \ldots, t_{m+1}$ inspection times.

probability that one working unit will fail during one step is equal to

$$p = \int_0^{\Delta} \lambda \exp(-\lambda t)\, dt. \tag{4.1}$$

The rule for switching on the standby elements could be written as follows:

$$s = f(k, i). \tag{4.2}$$

Here k denotes the number of inspection, $k = 0, \ldots, m$, i denotes the total number of nonfailed elements at the moment the inspection is performed and s denotes the total number of elements switched into operation for a time period (t_i, t_{i+1}), $1 \leq s \leq i$.

Now, suppose that a rule having the form (4.2) is chosen. Let us define the random process ξ_k, $k = 0, 1, \ldots, m+1$, as follows: $\xi_k = E_\infty$ if at the moment t_k of inspection the total failure is discovered. (It will be recalled that the total failure means that all elements which started to work at $t = t_{k-1}$ have failed up to time t_k.) If there is no total failure, $\xi_k = E_i$, where i is the total number of all nonfailed elements at the moment $t = t_k$, $i = 1, 2, \ldots, r$. The state E_∞ will be assumed to be an absorbing one. For $t = 0$ we set $\xi_0 = E_r$.

Example. Let $r = 4$ and the rule s have the following form:

$$f(0, i) = i - 2, \qquad i > 2.$$

That is, at the beginning ($t_0 = 0$) only two elements are operating.

$$f(1, i) = \begin{cases} i - 1, & i > 1 \\ 1, & i = 1 \end{cases}.$$

If at the moment $t = t_1$ there is only one good element, then it must be utilized; otherwise, all nonfailed elements except one must be engaged.

$$f(l, i) = i, \qquad l > 2.$$

This means switching on all nonfailed elements after all inspections are made at $t = t_l$, $l \geq 2$.

Figure 29 demonstrates one sample function of the process ξ_k when it was governed by the described rule $f(k, i)$; in the intervals (t_0, t_1), (t_1, t_2) one element failed and in the interval (t_2, t_3), two elements failed. The measure of this sample function is equal to

$$\pi_\xi = p_{43} p_{32} p_{2\infty},$$

where

$$p_{43} = p_{32} = \binom{2}{1} p(1-p), \qquad p_{2,\infty} = p^2. \quad \square$$

It is clear that under the above circumstances the process ξ_k is a

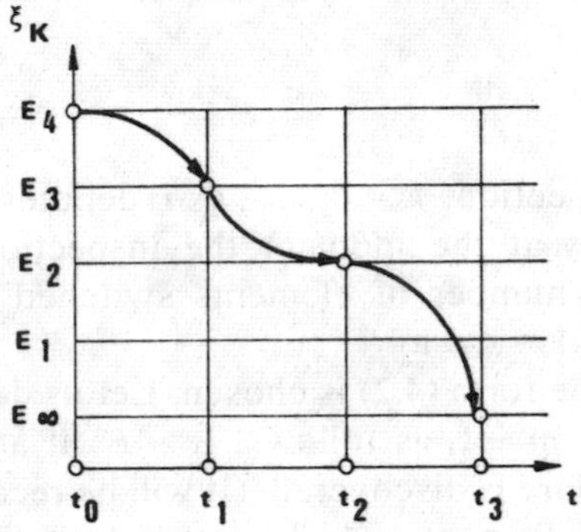

Figure 29. Sample function of the process ξ_k.

Markov chain. Denote

$$b(s, l) = \binom{s}{l} p^l (1-p)^{s-l}. \tag{4.3}$$

Thus, we have the following formulas for transition probabilities in that chain:

$$\begin{aligned} &P\{\xi_{k+1} = E_{i-l} | \xi_k = E_i\,;\ s = f(k, i)\} = p_{i,i-l}(s) = b(s, l), \\ &P\{\xi_{k+1} = E_\infty | E_k = E_i\,;\ s = f(k, i)\} = p_{i,\infty}(s) = b(s, s), \\ &P\{\xi_{k+1} = E_j \,|\, \xi_k = E_\infty;\ s = f(k, i)\} = \begin{cases} 0, & E_j \neq E_\infty; \\ 1, & E_j = E_\infty. \end{cases} \end{aligned} \tag{4.4}$$

Indeed, the transition $E_i \to E_{i-l}$ occurs if and only if exactly l of s working elements have failed during one step.

Now, without explanation, costs r_{ij} will be introduced for transition $E_i \to E_j$ in process ξ_k. It will be shown later that our total average costs are equal to the probability of the system's failure.

We set

$$r_{ij} = \begin{cases} 0, & \text{if } E_i \neq E_\infty,\ E_j \neq E_\infty; \\ 1, & \text{if } E_i \neq E_\infty,\ E_j = E_\infty; \\ 0, & \text{if } E_i = E_\infty. \end{cases} \tag{4.5}$$

In other words, the only cost equal to 1 for the transition into the failure state is counted.

From (4.4) and (4.5) it follows that the one-step cost for state E_i is equal to

$$\omega_i(s) = \begin{cases} p_{i,\infty}(s), & \text{if } E_i \neq E_\infty; \\ 0, & \text{if } E_i = E_\infty. \end{cases} \tag{4.6}$$

Our goal is to find an optimal function $f(k, i)$ which would provide the minimal average total penalty over the interval $(0, t_{m+1})$. The situation we are concerned with now was described in general terms in Section 3, Chapter 1, as "finite-step process". The CA in that case is a choice of a number of elements switched on for one step. The function $s = f(k, i)$ plays here the role of the strategy governing the process ξ_k. Generally speaking, $s = f(k, i)$ is the strategy of the second type because the solution depends on the number of the step. Let us turn now to the evaluation of recurrence relations for the costs. It will be assumed that we know the optimal strategy for a process starting at time $t = t_1$ with i nonfailed elements and lasting up to time $t_{m+1} = \theta$. Let $V_i(m)$ denote the corresponding average cost. Consider now the $(m+1)$-step process starting at time $t = 0$ with j nonfailed elements. We begin with switching on k_j elements at time $t = 0$, $k_j \leq j$, and from time t_1 we shall follow the optimal strategy. This gives a total cost $\hat{V}_j(m+1)$ which consists of a one-step cost for the state $\xi_0 = E_j$, plus the further cost over the interval (t_1, t_{m+1}). Calculating the costs, we take into account that if at interval (t_0, t_1) z elements fail $(z < k_j)$, the further cost is equal, according to the definition, to $V_{j-z}(m)$. If k_j elements fail, the further cost should be equal to zero because a transition to the absorbing state E_∞ has occurred.

According to (4.4) and (4.6) we write:

$$\hat{V}_j(m+1) = b(k_j, k_j) + \sum_{z=0}^{k_j-1} b(k_j, z) V_{j-z}(m),$$
$$j = 1, 2, \ldots, r; \quad m = 0, 1, 2, \ldots; \quad V_j(0) = 0.$$

There remains now only to apply the principle of optimality:

$$V_j(m+1) = \min_{1 \leq k_j \leq j} \left[b(k_j, k_j) + \sum_{z=0}^{k_j-1} b(k_j, z) V_{j-z}(m) \right],$$
$$j = 1, 2, \ldots, r; \quad m = 0, 1, 2, \ldots; \quad V_j(0) = 0. \tag{4.7}$$

In order to find the optimal strategy the "backward motion" must be applied; that is, at first the values $V_j(1)$ must be found by means of $V_j(0) = 0$. Proceeding from these data, the values $V_j(2)$ should be calculated, and so on. The technique of this computation will be demonstrated later.

Now let us clear up the physical meaning of values $V_j(m)$. From (4.7) it follows that

$$V_j(1) = \min_{1 \le k_j \le 1} b(k_j, k_j) = \min_{1 \le k_j \le j} p^{k_j} = p^j = b(j, j).$$

This indicates that $V_j(1)$ is equal to the probability that all elements will fail in the interval (t_m, t_{m+1}).

Consider now the two-step process starting at time $t = t_{m-1}$ with j nonfailed elements. According to (4.7),

$$V_j(2) = \min_{1 \le k_j \le j} \left[b(k_j, k_j) + \sum_{z=0}^{k_j - 1} b(k_j, z) V_{j-z}(1) \right].$$

The total failure in a two-step process can occur in two ways: either all k_j elements switched at $t = t_{m-1}$ fail, or z elements fail $(z < k_j)$ at the first step and all the remaining $j - z$ element fail at the second step.

Now it is clear that the expression in brackets is equal to the probability of total failure in a two-step process, calculated under the following conditions:

(a) the process starts at time $t = 0$ with j nonfailed elements;
(b) the first CA is "switch k_j elements";
(c) the second step is performed optimally according to the circumstances at the moment $t = t_m$.

This means that the value $V_j(2)$ is equal to a minimal probability of a failure in a two-step process. By analogy, it can be shown that $V_j(m)$ gives the desired value of a minimal probability of a failure in a m-step process.

The above reasoning demonstrates the fact that if one-step costs have the form (4.5), then the total average cost in a m-step process is equal to the probability of a total failure occurring during one of steps $k = 1, 2, \ldots, m$ (see Section 3.4 of Chapter 1).

4.3. Example

A system having $r = 3$ elements is considered on a time interval $(t_0 = 0, t_5)$. Probability of a failure during one step for an element is $p = 0.1$. The optimal strategy $s = f(k, j)$ has to be found.

It is obvious that $s = f(4, j) = j$ in a one-step process starting at t_4 with j

nonfailed elements ($j = 1, 2, 3$). The probabilities of the failure are

$$V_1(1) = p = 0.1; \qquad V_2(1) = p^2 = 0.01; \qquad V_3(1) = p^3 = 0.001.$$

Now consider a two-step process starting at $t = t_3$. Applying formula (4.7) we get

$$V_j(2) = \min_{1 \le k_j \le j} \left[b(k_j, k_j) + \sum_{z=0}^{k_j - 1} b(k_j, z) V_{j-z}(1) \right].$$

Substituting $j = 1$ one obtains by using (4.3)

$$V_1(2) = p + (1-p) V_1(1) = 0.190.$$

In a similar manner, (4.7) gives

$$V_2(2) = \min \begin{cases} p + (1-p) V_2(1) = 0.1 + 0.9 \cdot 0.01 = 0.109, & k_j = 1; \\ p^2 + (1-p)^2 V_2(1) + 2p(1-p) V_1(1) & \\ = 0.01 + 0.81 \cdot 0.01 + 2 \cdot 01 \cdot 0.9 \cdot 0.1 = 0.0361, & k_j = 2. \leftarrow \end{cases}$$

$$V_3(2) = \min \begin{cases} p + (1-p) V_3(1) = 0.1 + 0.9 \cdot 0.001 = 0.101, & k_j = 1; \\ p^2 + (1-p)^2 V_3(1) + 2(1-p) p V_2(1) & \\ = 0.01 + 0.81 \cdot 0.001 + 2 \cdot 0.9 \cdot 0.1 \cdot 0.01 = 0.013, & k_j = 2; \\ p^3 + (1-p)^3 V_3(1) + 3(1-p)^2 p V_2(1) & \\ + 3(1-p) p^2 V_1(1) & \\ = 0.001 + 0.729 \cdot 0.001 + 3 \cdot 0.81 \cdot 0.1 \cdot 0.01 & \\ + 3 \cdot 0.9 \cdot 0.01 \cdot 0.1 = 0.00686, & k_j = 3. \leftarrow \end{cases}$$

Therefore, $V_1(2) = 0.190$, $V_2(2) = 0.0361$, $V_3(2) = 0.00686$. Here the optimal strategy also indicates that all nonfailed units must be put into operation:

$$s = f(3, j) = j, \qquad j = 1, 2, 3.$$

Similarly, we will find that for the three-step process

$$V_1(3) = 0.271, \qquad V_2(3) = 0.0734, \qquad V_3(3) = 0.0198.$$

Again, the optimal rule consists of engaging all nonfailed elements: $s = f(3, j) = j_2$, at $t = t_2$.

Let us turn to a four-step process starting at $t = t_1$, with $V_1(4) = 0.144$, $V_2(4) = 0.118$. Here $s = f(1, 1) = 1$, $s = f(1, 2) = 2$. The situation changes for $j = 3$:

$$V_3(4) = \begin{cases} 0.118; & k_j = 1, \\ 0.0392; & k_j = 2, \leftarrow \\ 0.0406; & k_j = 3. \end{cases}$$

Finally, for the five-step process which starts at $t = 0$ with three elements we have:

$$V_3(5) = \begin{cases} 0.063; & k_j = 2, \leftarrow \\ 0.135; & k_j = 1, \\ 0.0675; & k_j = 3. \end{cases}$$

Here again one element must be kept as a spare, and the two others must be working. The probability of system's failure will be equal to 0.063. For a comparison it is worth remarking that the strategy "always to switch on all nonfailed elements" gives the failure probability $(1-(1-p)^5)^3 = 0.069$, that is, 9% greater.

4.4. *Optimality of "Zero-strategy"*

In many examples we see that the optimal strategy has the following simple structure: at every step of the process *all* nonfailed elements are working. Let us call this a "zero-spare-strategy" or simply a "zero-strategy". Sufficient conditions providing the optimality of a "zero-strategy" were given in [37]. Reference [65] contains the following necessary and sufficient conditions for the optimality of the "zero-strategy".

In a n-step process which starts with r nonfailed elements, "zero-strategy" is optimal if and only if the following inequality holds:

$$1-(1-p)^n \leqslant (1-p)^{(r-2)/(r-1)},$$

where p is defined by (4.1).

4.5. *Possible Generalizations*

The system with renewal of failed elements. Up to this point the assumption has been made that failed elements are not renewable. One might consider a situation where the failed elements are repaired (renewed) during the interval between inspections. It will be assumed that the renewal time for every failed element has an exponential distribution and therefore, the probability of a renewal during one step is a constant value, e.g. equal to q. Now the decision about the number of elements switched into the work depends on the total number of nonfailed elements, the number of the step and on the number of elements in repair:

$$s = f(k, i, n).$$

s is the number of elements which must be engaged, i is the number of inspection and n is the number of failed standby elements in repair.

Reasoning very similar to that given above leads to recurrence relations for this case (see [37] and [38]).

We now turn to a system with many kinds of elements where it is possible to make *a redistribution of nonfailed units.*

Let us imagine a system which has two types of standby elements which are mutually not interchangeable. Figure 30 shows such a system

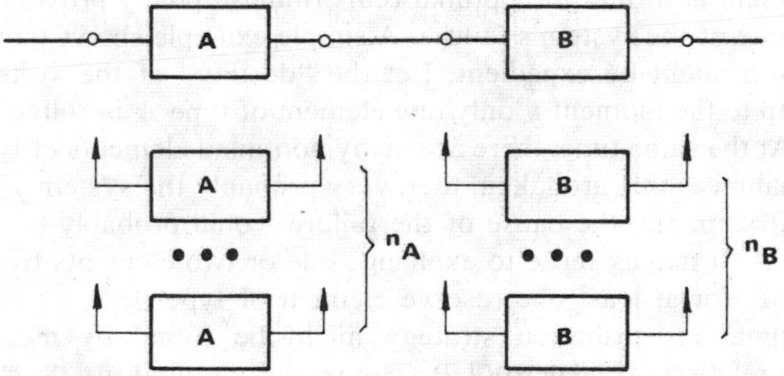

Figure 30. A system with two different kinds of "cold" standby units.

where both types of elements are working in the "cold" standby. The following problem is often solved in reliability theory.

It is necessary to find numbers n_A and n_B of the elements which provide the minimal probability of a system's failure over a given time interval $(0, \theta)$. Failure occurs if all elements of type A or B or both have failed during the period $(0, \theta)$. A reasonable statement of the problem, of course, must provide some constraints on the number of elements. For example, the following constraint must be kept:

$$c_A n_A + c_B n_B \leqslant C.$$

Coefficients c_A, c_B are costs of elements and the value C is the total cost.

Now it will be assumed that some time moments $t_1, t_2, \ldots, t_m$ are given during which it is possible to *redistribute* the number of standby elements between two subsystems A and B. Assume that the inspection is made at time t_i. Up to then, there have been only n'_A and n'_B nonfailed elements of types A and B respectively. The total value (cost) of all nonfailed elements is equal, therefore, to

$$C' = n'_A C_A + n'_B C_B.$$

Now the redistribution is allowed between the subsystems A and B with the possibility to choose new numbers n''_A and n''_B of standby elements for both subsystems by preserving the following condition:

$$n''_A c_A + n''_B c_B \leqslant C'.$$

Such a procedure might be performed after every inspection made at time t_i, $i = 1, \ldots, m$.

The problem is to find the optimal redistribution policy providing the minimal value of the system's failure. A simple example shows that such redistribution might be expedient. Let the "destiny" of the system be such that up to the moment t_i only one element of type A is "alive", that is $n'_A = 0$. At the same time, there are many nonfailed elements of type B. If no special measures are taken, then very probably the system will fail on the next step, and the cause of the failure would probably be in the subsystem A. It makes sense to exchange one or two elements from the subsystem B for at least one reserve element of type A.

The optimal redistribution strategy might be found by means of recurrence relations similar to (4.7). The reader can find in [38] a more detailed discussion of that problem.

5. *Optimal Replacement Policy for a "Multiline" System*

5.1. *Statement of the Problem*

In this section, a system consisting of n identical subsystems will be considered. Every subsystem works and fails independently of other subsystems, while the replacement (repair) of the failed subsystem requires that the work of all systems be stopped. Such a system is called a "multiline" and its subsystems, "lines" [56, Section 17].

It will be assumed that the life span of every line has a known DF $F(t)$ and that there is an immediate signalling of failure.

The servicing of the systems proceeds in the following way. When a total number of failed lines is equal to the prescribed number k, $1 \leqslant k \leqslant n$, the system is stopped and the failed lines are replaced by new ones (repaired). After the replacement, the system works as if brand new. Figure 35 shows an operation diagram of a multiline system with $n = 6$ lines. Its repair starts after $k = 3$ lines fail.

Now assume that the system's return is proportional to the total failure-free operation time of all lines. Thus, for the period (t_0, t_0') (see Figure 31), the total efficiency is proportional to the value $t_1 + t_2 + 4t_3$.

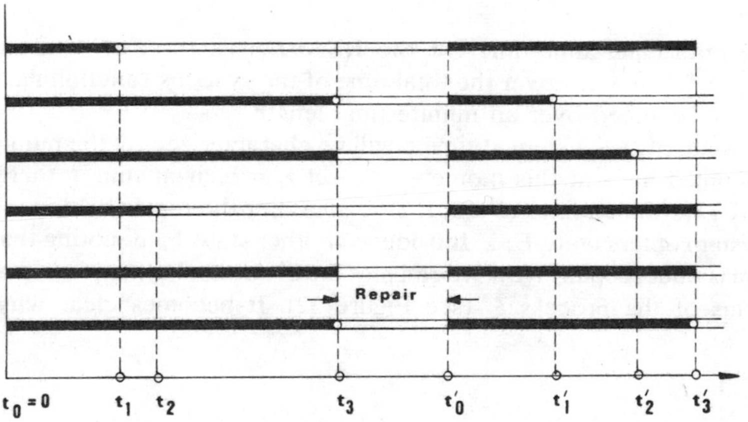

Figure 31. A diagram of multiline system operation; t_1, t_2, t_3, t_1', t_2', t_3' instants when the lines failed.

The time for making repairs when k lines have failed is assumed to be equal to $t(k)$. This time may also be a random value having ME $\overline{t(k)}$. The problem is to find such value k which provides the maximal efficiency per unit time. The solution is not evident: if the value $k = 1$ is chosen, the system would be repaired very often; if on the contrary, the value is maximal ($k = n$), repairs would occur infrequently, but for a large part of time the multiline system would work with low efficiency.

The prototype for a multiline system is a special kind of automatic production line, called a "rotor"-type automatic production line (RTAPL)

([54] and [56]). Such lines are in wide use when the output is very large. Usually, RTAPL consists of a series of special devices called rotors adjusted to carrying out one technological operation. The transferring of half-finished articles from one rotor to the next is performed by special transporting devices. Often all rotors of the RTAPL have equal number of positions, and therefore, the corresponding positions form one separate line. Usually the construction of RTAPL provides the possibility that when a failure occurs, the failed subsystem can be switched out without stopping the entire RTAPL. The efficiency of the RTAPL is proportional to its output or to the total amount of operation time of all its lines.

5.2. *The Relation for the Expected Return per Unit Time when Every Line Has a Constant Failure Rate*

The return per unit time for the RTAPL is a ratio of total active worktime for all lines over the total time of the systems functioning. This ratio will be taken over an infinite time length.

The state of the system at time t will be characterized by the number ξ_t of nonfailed lines at this moment. We set $\xi_t = E_i$ if at time t there are exactly i nonfailed lines ($i = 0, 1, 2, \ldots, n$). Let the repair be done when the system enters state E_{n-k}. Introduce another state E_p denoting that the system is under repair. Now we can plot the diagram of system's operation in terms of the process ξ_t (see Figure 32). It becomes clear why the

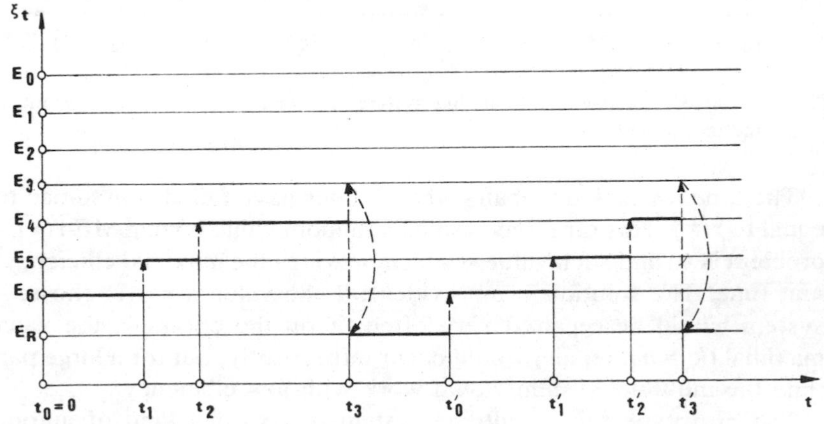

Figure 32. A sample function of the process ξ_t.

problem of finding the optimal replacement rule for the RTAPL is considered in this chapter: for the decision concerning replacement, information about the actual state of the random process ξ_t must be taken into account. The control of the process ξ_t here is a "shift" of the process from some "critical" state E_{n-k} into state E_R.

The time length between two subsequent switchings into the operation will be called a cycle. Thus, in Figure 32, one whole cycle having the length t_0' is represented.

Now the returns are defined. The return $i \cdot \theta_i$ is accumulated when the system spends time θ_i in state E_i. Let τ_j denote the time when the failure of the jth line occurs. Thus, the process ξ_t stays in state E_n the random time τ_1, in state E_{n-1} the random time $\tau_2 - \tau_1$, in state E_i the random time $\tau_{n-i+1} - \tau_{n-i}$. Therefore, the total return $W(k)$ for one cycle is equal to

$$W(k) = \tau_1 n + (\tau_2 - \tau_1)(n-1) + (\tau_3 - \tau_2)(n-2) + \cdots + (\tau_k - \tau_{k-1})(n-k+1) = \tau_1 + \tau_2 + \cdots + \tau_k + \tau_k(n-k). \tag{5.1}$$

This gives us the expected return in one cycle:

$$M\{W(k)\} = \sum_{j=1}^{k} M\{\tau_j\} + M\{\tau_k\}(n-k).$$

The ME of the cycle's length is equal to

$$M\{\tau_k + t(k)\} = M\{\tau_k\} + \tau(k). \tag{5.2}$$

Now it is clear that the expected return per unit time is

$$v(k) = \frac{M\{W(k)\}}{M\{\tau_k\} + \bar{t}(k)}. \tag{5.3}$$

It is convenient to characterize the efficiency of RTAPL by means of the ratio

$$\chi_k = v(k)/n. \tag{5.4}$$

The most difficult point in the whole calculation is finding the values $M\{\tau_i\}$. Consider first the case where all lines have a constant failure rate. Such a supposition is realistic when one individual line consists of many parts and might be considered as a "large" system. Such a system has usually a constant failure rate. If one line has a failure rate λ, then the failure rate for the state E_n is λn or, in other words, the random variable τ_1 has an exponential distribution of time-to-failure with the parameter λn

and

$$M\{\tau_1\} = 1/\lambda n.$$

After the first failure, the failure rate will be equal to $\lambda(n-1)$ and, therefore $M\{\tau_2 - \tau_1\} = 1/\lambda(n-1)$. In the same way we obtain that

$$M\{\tau_i\} = \frac{1}{\lambda}\left[\frac{1}{n} + \cdots + \frac{1}{n-i+1}\right]. \tag{5.5}$$

Substituting this expression into (5.3), we get the following simple formula for the efficiency:

$$\chi_k = \frac{1}{n}\frac{k}{\sum_{i=1}^{k}\frac{1}{n+1-i} + \bar{t}(k)\lambda}. \tag{5.6}$$

It is worthwhile mentioning that in the case of constant failure rate, the repair of failed lines renews the whole system entirely.

Investigating the optimal value of k which would maximize the efficiency, we shall assume that the repair time has the following form:

$$\overline{t(k)} = N + Lk. \tag{5.7}$$

Consider first the case $N = 0$. (5.6) gives

$$\chi_k = \frac{1}{n} \cdot \frac{1}{\lambda L + \frac{1}{k}\sum_{i=1}^{k}(n+1-i)^{-1}}. \tag{5.8}$$

It is easy to demonstrate that the second term in the denominator increases when k increases. Therefore, in the case $N = 0$, the optimal $k = 1$ or, in other words, the RTAPL must be stopped for repair after the first failure occurs.

Now assume that the repair time does not depend on the number of failed lines, that is, $L = 0$. Now, usually the optimal k is not equal to one. Figure 33 shows a number of curves for the efficiency depending on the number k. The upper index at χ denotes the number of lines in the system. It is interesting to note that the optimal k increases with n and N.

For the case $N > 0$, $L > 0$, the optimal k might be found by means of simple calculations, using formula (5.6).

Remarks. The problem of finding the optimal service strategy was solved without applying the theory of controlled SMP, owing to the simple form of the expression of average return per unit time. Neverthe-

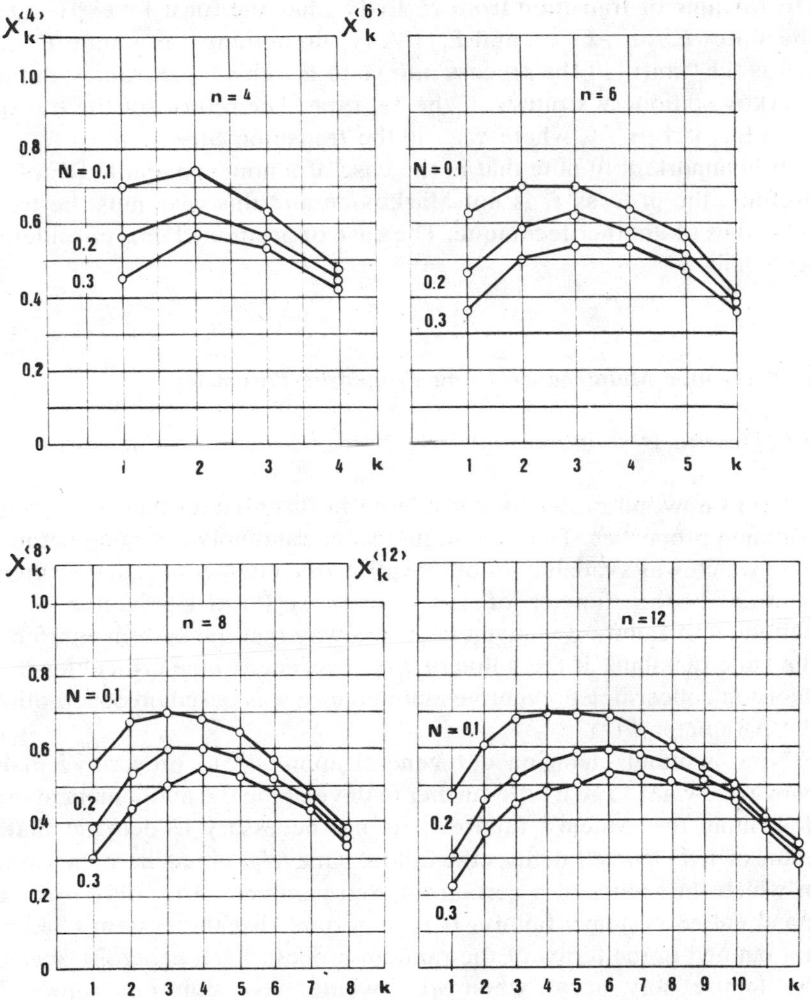

Figure 33. Dependence of the efficiency of one line in the n-line system on the "stopping" parameter k ($\lambda = 1$).

less, it is interesting to mention what form the solution would have in terms of controlled SMP. Let the repair rule (the number k) be chosen. Then by the assumption of constant failure rate for lines, the process ξ_t is a SMP. The transitions in this process follows the scheme:

$$E_n \to E_{n-1} \to \cdots E_{n-k} \to E_R \to E_n \to \cdots.$$

DF for time of transition from E_i to E_{i-1} has the form $1-\exp(-\lambda_i t)$. In the states $E_n, \ldots, E_{n-k+1}$ and E_p, CA is "do nothing". For state E_{n-k}, the CA is the "shift" of the process into state E_p. This repair rule is, in fact, a Markov stationary strategy of the 1st type. The return for the transition $E_i \to E_{i-1}$ is $i \cdot \tau_{i,i-1}$, where $\tau_{i,i-1}$ is the transition time.

It is important to note that in the case of a nonexponential DF of lines lifetime, the process ξ_t is not Markovian and this case must be treated by means of another technique. The case of arbitrary DF is considered in Appendix 3.

6. Preventive Maintenance Using Prognostic Parameter

6.1. General Description and Definition of Prognostic Parameter

Up to now, all problems considered in this chapter had the following common properties. The system had some randomly changing parameter $\eta(t)$ which was available for observation (the number of failed elements – Section 2; the number of failed lines in RTAPL – Section 5). The definition of failure was given in such a way that the system entered into the state of failure if the value of $\eta(t)$ exceeded some critical level. The decision concerning preventive maintenance was based on observation of the parameter $\eta(t)$.

Now, without changing our general approach to preventive maintenance, we want to go a little further in developing the mathematical model describing the system's failure. It is not necessary to demand that the value of $\eta(t)$ in *itself* defines the failure. One might imagine such a system in which the values of a certain random parameter $\eta(t)$ only *predict* an event called system's failure. It is important that the system's failure is not defined in the terms of the random process $\eta(t)$; generally speaking, the failure may occur when $\eta(t)$ would have different values. The postulate is only a stochastic dependence between the probability of the failure in the interval $(t, t+\Delta t)$ and the actual value of the parameter η at moment t. For example, the dependence might be as follows: the increase of $\eta(t)$ leads to an increase of failure probability in the interval $(t, t+\Delta t)$. Such a model of failure was considered in [62] and [41, Chapter 4]. The random parameter $\eta(t)$ having the above-mentioned property will be called the prognostic parameter.

Prognostic parameters usually are inherent in situations where the failure is preceded by a process of accumulation of damages and is itself a rapid change in the system influenced by this process.

In mechanical systems, the failure (the breakage of some part) may be caused by the impact of a "peak-load". The amount of accumulated wear or fatigue damages plays the role of the prognostic parameter. The more extensive these damages are, the lower the "resistance abilities" of the system and the higher the probability of a failure when a "peak-load" appears.

There are some instant failures occurring in electronic equipment (such as a short circuit, dielectric breakdown, breaking of a circuit, etc.) which are speeded up by some gradual process of deterioration, e.g. corrosion, mechanical deformation, chemical reactions and so on. These processes play a prognostic role.

In a system with so-called "hot" standby, the load on the main unit may depend on the number of working units in the standby. The more units that fail, the greater is the probability of a failure in the main unit. The number of failed units here plays the role of the prognostic parameter.

We come now to the formal definition of a prognostic parameter, after which we will formulate the problem of optimal preventive maintenance using this parameter.

It is assumed that the system has a randomly changing parameter $\eta(t)$. An event "failure" is defined. Let t denote time, τ the random life span of the system (time-to-failure), η' be a certain "history" of the parameter $\eta(t)$ on the time interval $(0, t)$. The parameter $\eta(t)$ is called a *prognostic* if the following property takes place:

$$\underset{\Delta t \to 0}{\boldsymbol{P}} \{t < \tau \leq t + \Delta t \mid \eta', \eta(t) = \eta, \tau > t\}$$
$$= \boldsymbol{P}\{t < \tau \leq t + \Delta t \mid \eta(t) = \eta, \tau > t\} = \lambda(\eta)\Delta t + o(\Delta t). \quad (6.1)$$

This property means that the probability of a failure in the interval $(t, t + \Delta t)$, under the condition that there was a certain history of the process $\eta(t)$ before the failure and that the system had not failed earlier, is equal to the value Δt multiplied by a function of the actual value of the parameter η. Physically, this means that only the actual level of $\eta(t)$ influences the failure probability in a future short time period.

The *time* elapsed since the last failure also may be considered as a prognostic parameter. Indeed, if $\eta(t) = t$, the relation (6.1) is equivalent to the following one:

$$\boldsymbol{P}\{t < \tau \leq t + \Delta t \mid \tau > t\} = \lambda(t)\Delta t + o(\Delta t). \quad (6.2)$$

In this formula, $\lambda(t)$ is the well-known failure rate calculated by means of the expression

$$\lambda(t) = F'(t)/(1 - F(t)),$$

where $F(t)$ is a differentiable DF.

In the general case, one might consider the value $\lambda(\eta)$ as a generalized failure rate computed for the "time" η. If the system has an observable prognostic parameter, then the preventive maintenance can be organized in the following way. A certain "critical" level h for the parameter $\eta(t)$ is chosen. If the parameter reaches that value before the failure occurs, then a PR is performed; if the failure occurs before the prognostic parameter reaches the critical value, an ER is performed.

6.2. Formula for Operational Readiness when Preventive Maintenance Uses the Prognostic Parameter

It will be assumed that PR and ER entirely renew the system. After ER or PR the prognostic parameter "is replaced" and it takes the value $\eta(0) = \eta_0$. The ER lasts time $t_a = 1$, the PR lasts time $t_p(h)$, $t_p(h) < 1$. Figure 34 illustrates the system's performance using a prognostic parameter.

We note if $\eta(t)$ is the time elapsed since the last failure, then the critical value h is simply the critical age and there is no difference between this

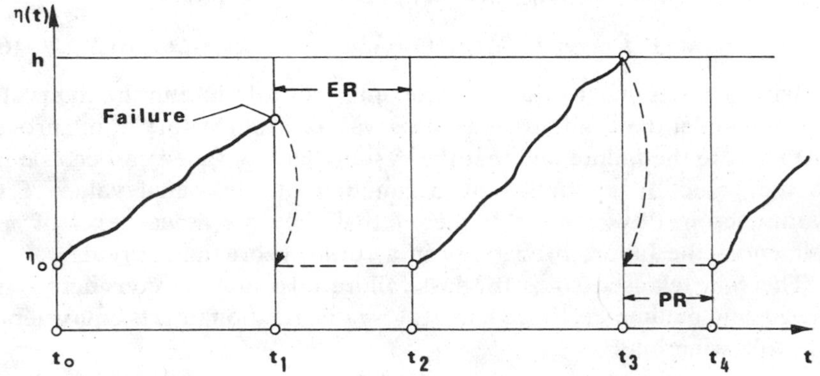

Figure 34. Scheme of preventive maintenance using prognostic parameter $\eta(t)$.

scheme of preventive maintenance and the well-known age replacement (Sections 2 and 3 in Chapter 2).

The system is assumed to function for a long time period and the problem is to find the critical level h which maximizes the operational readiness.

First the expression for the OR under the constant level h will be obtained. Let us define three states in the system:

E_1 – the system is operating;
E_2 – the ER is carried out;
E_3 – the PR is carried out.

It is not difficult to show that the random process $\xi_t = E_i$, $i = 1, 2, 3$, is a SMP.

The matrices of transition probabilities $\mathbf{P}(h)$ and transition distributions $\mathbf{F}(h)$ have the following general form:

$$\mathbf{P} = \mathbf{P}(h) = \begin{Vmatrix} 0 & p_{12}(h) & p_{13}(h) \\ 1 & 0 & 0 \\ 1 & 0 & 0 \end{Vmatrix},$$

$$\mathbf{F} = \mathbf{F}(h) = \begin{Vmatrix} 0 & F_{12}(t;h) & F_{13}(t;h) \\ G_a(t) & 0 & 0 \\ G_p(t) & 0 & 0 \end{Vmatrix}. \tag{6.3}$$

Here $G_a(t)$ and $G_p(t)$ are DF of ER and PR respectively. Denote the expected return for the transition $E_1 \to E_2$ and $E_1 \to E_3$ equal to the time of those transitions. Returns for transitions into state E_1, $E_2 \to E_1$ and $E_3 \to E_1$, are assumed to be zero. The one-step return for state E_1 is

$$\omega_1(h) = p_{12}(h) \int_0^\infty t \,\mathrm{d}F_{12}(t;h) + p_{13}(h) \int_0^\infty t \,\mathrm{d}F_{13}(t;h) = \int_0^\infty t \,\mathrm{d}B(t;h), \tag{6.4}$$

where

$$B(t;h) = p_{12}(h)F_{12}(t;h) + p_{13}(h)F_{13}(t;h). \tag{6.5}$$

It is obvious that $B(t;h)$ is the probability that the process ξ_t will stay in state E_1 not longer than time t.

According to the definition $\omega_2 = \omega_3 = 0$; the expected return per unit time is equal to the quotient:

$$g(h)=\frac{\omega_1(h)}{\nu_1(h)+p_{12}(h)\nu_2(h)+p_{13}(h)\nu_3(h)}, \tag{6.6}$$

where

$$\nu_1(h)=\omega_1(h), \qquad \nu_2(h)=1, \qquad \nu_3(h)=t_p(h). \tag{6.7}$$

This follows from the direct application of formula (3.44), Chapter 1. The meaning of the expression (6.6) is simple: the numerator is the average length of the active work period, the denominator is equal to the average length of a cycle "work-repair". This expression is similar to formula (2.12) of Chapter 2, which gives the value of OR for the age replacement.

It is obvious that (6.6) does not lead immediately to finding the optimal critical level because it is not clear how the values $\omega_1(h)$, $p_{12}(h)$, $p_{13}(h)$ depend on the behaviour of the prognostic parameter and on the function $\lambda(\eta)$. We shall consider in detail three cases: the prognostic parameter is a finite-state Markov process, a Poisson process and a linear random process.

6.3. The Case where the Prognostic Parameter is a Markov Process with Finite Number of States

Let $L=\{L_0, L_1, \ldots, L_n, L^*_{n+1}\}$ be the set of states of a time-homogeneous Markov process φ_t defined by means of transition rates a_{ij}, $i, j=0, \ldots, n+1$, $i\neq j$, $0\leqslant a_{ij}<\infty$. The state L_0 is an initial one: $P\{\xi_0=L_0\}=1$.

The state L^*_{n+1} is a unique absorbing state. Entering that state means the system has failed. We recall that the values a_{ij} have the following meaning:

$$\underset{\Delta t\to 0}{P}\{\varphi_{t+\Delta t}=L_j\,|\,\varphi_t=L_i\}=a_{ij}\Delta t+o(\Delta t), \qquad i=0,\ldots,n; \quad j\neq i;$$

$$\underset{\Delta t\to 0}{P}\{\varphi_{t+\Delta t}=L_i\,|\,\varphi_t=L_i\}=1-\sum_{j\neq i}a_{ij}\Delta t+o(\Delta t), \qquad i=0,\ldots,n. \tag{6.8}$$

The values $a_{i,n+1}$ indicates the rate of transition from E_i into the failure state:

$$\underset{\Delta t\to 0}{P}\{\varphi_{t+\Delta t}=L^*_{n+1}\,|\,\varphi_t=L_i\}=a_{i,n+1}\Delta t+o(\Delta t). \tag{6.9}$$

Comparing this expression with (6.1), one could conclude that the number of state i ($i = 0, 1, \ldots, n$) determines the probability of the failure in the interval $(t, t + \Delta t)$ and, therefore, φ_t might be considered as a prognostic parameter. Let the state $L_i = L_h$ be a critical one and organize the preventive maintenance as described in Section 6.1.

Our goal is to obtain the expression for $g(h)$. In order to do that, we shall modify the process φ_t by introducing one new absorbing state, namely, L_h. The new Markov process will be denoted φ_t^*. The transition rates for φ_t^* are the same as for φ_t except a_{hj} $(h \neq j)$, which must be equal to zero.

We had to introduce a new absorbing state L_h because one cycle of the system's work will be finished either if the process φ_t^* reaches the level L_h before the failure occurs or when the failure occurs before process φ_t^* reaches state L_h. Therefore, the trajectory of process φ_t^* breaks off by entering both states L_{n+1}^* and L_h.

Now the standard method will be applied for finding the values in (6.6).

Let $P_i(t)$ be the probability that the process φ_t^* is in state E_i at time t:

$$P_i(t) = \mathbf{P}\{\varphi_t^* = L_i \,|\, \varphi_0^* = L_0\}.$$

Let us form the matrix $\mathbf{A}$ having elements a_{ij} for $i \neq j$ and $a_{ii}^* = -\Sigma_{j \neq i}\, a_{ij}$ (see Section 3 of this chapter). The rows corresponding to states L_h and L_{n+1}^* will contain only zeros. For making further calculations, the following relation will be used:

$$\mathbf{P}'(t) = \mathbf{A}^T \mathbf{P}(t), \tag{6.10}$$

where $\mathbf{A}^T$ is the transposition of the matrix $\mathbf{A}$, $\mathbf{P}'(t)$ and $\mathbf{P}(t)$ are vectors consisting of elements $P_i'(t)$, $P_i(t)$, $i = 0, \ldots, n + 1$.

Now it should be noted that $p_{12}(h)$ in (6.6) is, in terms of the process φ_t^*, the probability that the trajectory will be absorbed by the state L_{n+1}^*. Similarly, $p_{13}(h)$ is the probability of an absorption by the state L_h. Therefore,

$$\begin{aligned} p_{12}(h) &= \lim_{t \to \infty} P_{n+1}(t) = P_{n+1}(\infty), \\ p_{13}(h) &= \lim_{t \to \infty} P_h(t) = P_h(\infty). \end{aligned} \tag{6.11}$$

Applying to both sides of (6.10) the Laplace transform

$$\pi_i(s) = \int_0^\infty e^{-st} P_i(t)\, dt, \tag{6.12}$$

we obtain the following system of linear algebraic equations:

$$(\mathbf{I}s - \mathbf{A}^T)\Pi(s) = \mathbf{P}(0), \tag{6.13}$$

where $\Pi(s)$ is a vector consisting of elements $\pi_i(s)$, $\mathbf{I}$ – the unit matrix, $\mathbf{P}(0) = \|1, 0, \ldots, 0\|^T$.

In order to find $p_{12}(h)$, $p_{13}(h)$, the following property of the Laplace transform will be used:

$$\lim_{t\to\infty} P_{n+1}(t) = p_{12}(h) = \lim_{s\to 0}\,[s\pi_{n+1}(s)]. \tag{6.14}$$

It is clear that

$$p_{13}(h) = 1 - p_{12}(h). \tag{6.15}$$

Therefore, in order to find $p_{12}(h)$ there is no necessity to calculate the originals for $\pi_i(s)$.

It should be noted also that the value $\omega_1(h) = \nu_1(h)$ in terms of the process φ_t^* is equal to the mathematical expectation of the walking time up to entering one of absorbing states.

To find this value there is no need to solve the system (6.10). We give an alternative method proposed in [58].

Let $\mathbf{A}_0 = \|a_{ij}^0\|$ denote the matrix of order n obtained from $\mathbf{A}^T$ by cancelling out two columns and two rows corresponding to the absorbing states (the $(h+1)$th and the last rows and columns). Now replace the elements of the first row in $\mathbf{A}_0$ by units and denote the new matrix by $\mathbf{A}_1$. It turns out that the mathematical expectation of the walking time for the process φ_t^* is equal to

$$\nu_1(h) = -\frac{\det\|\mathbf{A}_1\|}{\det\|\mathbf{A}_0\|}. \tag{6.16}$$

Now all values in formula (6.6) are known. We consider next an example which has no practical importance but demonstrates the calculation procedure.

6.4. *Example*

Consider a system with a randomly changing parameter $\eta(t)$. The parameter has four levels L_0, L_1, L_2 and L_3, and its evolution is monotone. The transition rates λ_i are known for the process $\eta(t)$ (see Figure 35a). It will be assumed that $\boldsymbol{P}\{\eta(0) = L_0\} = 1$. A certain failure

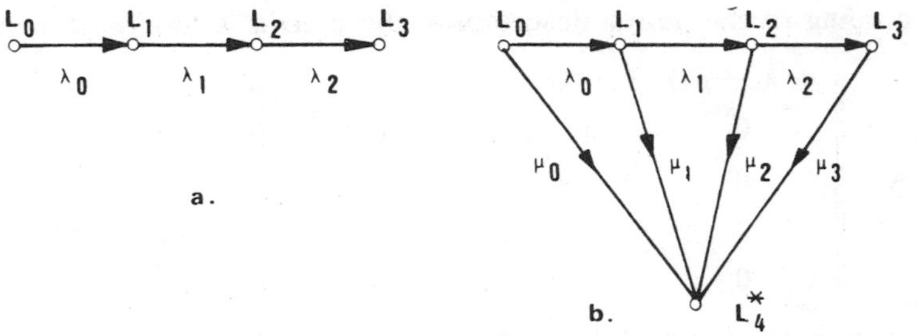

Figure 35. (a) The evolution of the prognostic parameter. (b) Transition rate diagram for the process ξ_t.

rate μ_i, $i = 0, 1, 2, 3$, corresponds to each level L_i and the values μ_i increase with i. Therefore, $\eta(t)$ might be considered as a prognostic parameter.

The state of the system, i.e., state of the parameter and appearance of the failure, might be described by means of a random process φ_t having states L_0, L_1, L_2, L_3 and L_4^*. The last one corresponds to the system's failure. It is clear that the process φ_t is a Markovian, time-homogeneous process. Its graph is plotted in Figure 35b.

The preventive maintenance is organized as follows. One of the states L_h, $h = 1, 2$ or 3, is declared as a marginal one. If the system enters that state (before the failure) a PR is carried out and the penalty $c(h)$ is paid. If the system fails before entering state L_h, an ER is made and the penalty c_a is paid. Our goal is to find the optimal value h which provides the minimal expected cost per unit time. The time for performing repairs is assumed to be negligible.

The expression for the average cost per unit time may be derived similarly to that described earlier for OR (see Section 6.2).

The average penalty is equal to the ratio

$$u(h) = \frac{p_{12}^*(h)c_a + p_{13}^*(h)c(h)}{\nu_1(h)}, \tag{6.17}$$

where $p_{12}^*(h)$ is the probability of the failure before $\eta(t)$ reaches the critical level, $p_{13}^*(h) = 1 - p_{12}^*(h)$. Indeed, the numerator gives the expected value of the penalty during one cycle and the denominator is equal to the average cycle length.

Assume that the state L_3 is a marginal one. We show how to calculate the value $u(h)$ when $h = 3$.

According to the above description, the matrix **A** has the form:

$$\mathbf{A} = \begin{Vmatrix} -(\lambda_0+\mu_0) & \lambda_0 & 0 & 0 & \mu_0 \\ 0 & -(\lambda_1+\mu_1) & \lambda_1 & 0 & \mu_1 \\ 0 & 0 & -(\lambda_2+\mu_2) & \lambda_2 & \mu_2 \\ 0 & 0 & 0 & 0 & 0 \\ 0 & 0 & 0 & 0 & 0 \end{Vmatrix}.$$

(Note that states L_3 and L_4^* are absorbing ones.) Cancelling two last rows and columns, we obtain the matrix $\mathbf{A}_0$:

$$\mathbf{A}_0 = \begin{Vmatrix} -(\lambda_0+\mu_0) & 0 & 0 \\ \lambda_0 & -(\lambda_1+\mu_1) & 0 \\ 0 & \lambda_1 & -(\lambda_2+\mu_2) \end{Vmatrix}.$$

The matrix $\mathbf{A}_1$ has the form

$$\mathbf{A}_1 = \begin{Vmatrix} 1 & 1 & 1 \\ \lambda_0 & -(\lambda_1+\mu_1) & 0 \\ 0 & \lambda_1 & -(\lambda_2+\mu_2) \end{Vmatrix}.$$

(6.16) gives

$$\nu_1(h=3) = \frac{-\det\|\mathbf{A}_1\|}{\det\|\mathbf{A}_0\|} = [(\lambda_2+\mu_2)(\lambda_0+\mu_1+\lambda_1)+\lambda_0\lambda_1]\prod_{i=0}^{2}(\lambda_i+\mu_i)^{-1}. \tag{6.18}$$

The system of equations (6.13) has the form

$$\begin{aligned} &\pi_0(s)(s+\lambda_0+\mu_0) && &&= 1,\\ -&\pi_0(s)\lambda_0 &&+\pi_1(s)(s+\lambda_1+\mu_1) &&= 0,\\ & &&+\pi_1(s)\lambda_1+\pi_2(s)(s+\lambda_2+\mu_2) &&= 0,\\ & &&-\pi_2(s)\lambda_2+\pi_3(s)s &&= 0,\\ -&\pi_0(s)\mu_0 &&-\pi_1(s)\mu_1-\pi_2(s)\mu_2 \qquad +\pi_4(s)s &&= 0. \end{aligned}$$

This can easily be solved. We find

$$\pi_3(s) = \frac{\lambda_0\lambda_1\lambda_2}{(s+\lambda_0+\mu_0)(s+\lambda_1+\mu_1)(s+\lambda_2+\mu_2)s}.$$

Using (6.14) and (6.15), we obtain:

$$\lim_{s\to 0}[s\pi_3(s)] = p_{13}^*(h=3) = \prod_{i=0}^{2}\frac{\lambda_i}{\lambda_i+\mu_i}.$$

Assume that the transition rates and penalties have the following values:

$$\lambda_0 = 1; \quad \lambda_1 = 2; \quad \lambda_2 = 3;$$

$$\mu_0 = 0.5; \quad \mu_1 = 1; \quad \mu_2 = 1.5; \quad \mu_3 = 6.$$

$$c_a = 1; \quad c(1) = 0.2; \quad c(2) = 0.25; \quad c(3) = 0.4.$$

After calculation for $h = 3$, we obtain $u(h = 3) = 0.83$.

By performing the same calculations for $h = 1, 2$, we find that $u(h = 1) = 0.70$ and $u(h = 2) = 0.75$. Therefore, the optimal critical level for the prognostic parameter is $L_h = L_1$. In other words, the optimal preventive maintenance is to carry out the PR immediately after the prognostic parameter enters state L_1. It is worth noting that if we ignore the presence of the prognostic parameter and perform only the ER when the failure occurs, we shall have a greater average penalty (approximately equal to 0.96). Therefore, in a given situation it is expedient to exploit the presence of the prognostic parameter.

It should be remarked here, concerning the critical state, that generally the role of the critical level L_h may be played by a set of states $\boldsymbol{L}_1$. If the system enters state $L' \in \boldsymbol{L}_1$ the PR should be carried out. That only slightly changes the calculation scheme and all states forming the set $\boldsymbol{L}_1$ must be considered as absorbing states.

6.5. The Poisson Process as a Prognostic Parameter

At this point we proceed with the case when the prognostic parameter $\eta(t)$ is a Poisson process having a constant rate λ. Let $\boldsymbol{E} = \{E_i, i \geqslant 0\}$ denote the set of states of this process and E^* the failure state. Assume that

$$\lambda(\eta) = \eta\mu + \mu_0 \tag{6.19}$$

(see formula (6.1)).

The model of occurrence of failure might be interpreted as follows. Shocks occur to the system with a constant rate λ. Every shock causes a constant amount of damage. When k shocks are accumulated up to the instant t, we say that the system is at time t in state E_k. Two changes can appear in the coming interval $(t, t + \Delta t)$: a new shock appears with probability $\lambda\Delta t + o(\Delta t)$ or the system enters failure state E^* with probability $(\mu_0 + k\mu)\Delta t + o(\Delta t)$. This model reflects a situation of a gradual weakening of the system caused by the process of accumulation of damages.

We introduce a random process $\eta^*(t)$ with states $\{E^*, E_0, E_1, \ldots,\}$. The state E^* will be an absorbing one. Let $P_k(t) = \boldsymbol{P}\{\eta^*(t) = E_k\}$, $k \geq 0$. By using routine technique, one can easily obtain the following set of differential equations for functions $P_k(t)$:

$$P_0'(t) = -(\lambda + \mu_0)P_0(t), \tag{6.20}$$
$$k \geq 1: \quad P_k'(t) = -\mu k P_k(t) - (\lambda + \mu_0)P_k(t) + \lambda P_{k-1}(t).$$

This set must be solved by the initial condition $P_0(0) = 1$, $P_k(0) = 0$, $k > 0$, which corresponds to the assumption that at starting point $t = 0$ the accumulated damage is equal to zero. The solution of (6.20) can be found in a recursive manner and it has the form:

$$k \geq 0: \quad P_k(t) = \mathrm{e}^{-(\lambda+\mu_0)t}(1 - \mathrm{e}^{-\mu t})^k \left(\frac{\lambda}{\mu}\right)^k \frac{1}{k!}. \tag{6.21}$$

The distribution function of failure-free operation time $\boldsymbol{\tau}$ might be written as follows:

$$\boldsymbol{P}\{\boldsymbol{\tau} \leq t\} = 1 - \sum_{k=0}^{\infty} P_k(t) = 1 - \exp\left[\frac{\lambda}{\mu} - (\lambda + \mu_0)t - \frac{\lambda}{\mu}\mathrm{e}^{-\mu t}\right]. \tag{6.22}$$

This DF has an increasing failure rate:

$$\rho(t) = \mu_0 + \lambda(1 - \mathrm{e}^{-\mu t}).$$

Now the state E_h will be selected as a critical one and preventive repairs will be performed when the system (formally – the process $\eta^*(t)$) enters this state. We are required to choose the number h in such a way as to minimize the expected cost per unit time. It is assumed that ERs and PRs are carried out instantaneously and the costs associated with them are 1 and C correspondingly.

The process $\eta^*(t)$ enters the critical state E_h in the interval $(t, t + \Delta t)$ if and only if the damage $\eta^*(t)$ accumulated up to time t was equal to E_{h-1} and the transition $E_{h-1} \to E_h$ occurred in this interval. Therefore, the probability of entering state E_h at some random time $\boldsymbol{\tau}$, $\boldsymbol{\tau} \leq t$, is equal to the integral

$$Q_{13}(t; h) = \int_0^t P_{h-1}(u)\lambda \, \mathrm{d}u. \tag{6.23}$$

Obviously,

$$p_{13}^*(h) = \lim_{t\to\infty} Q_{13}(t; h) \tag{6.24}$$

(see (6.3), the definition of $p_{13}^*(h)$). After some manipulations, we obtain the expression

$$p_{13}^*(h) = \lambda^h \left[\prod_{k=0}^{h-1} (\lambda + \mu_0 + k\mu)\right]^{-1}. \tag{6.25}$$

It has a transparent probabilistic meaning: $\lambda/(\lambda + \mu_0 + k\mu)$ is a conditional probability of jumping from E_k to E_{k+1} calculated under the condition that $\eta^*(t)$ has left state E_k.

The probability of the failure in the interval $(0, t)$ is equal to

$$Q_{12}(t; h) = \int_0^t \left[\sum_{k=0}^{h-1} P_k(u)(\mu_0 + k\mu)\right] du. \tag{6.26}$$

One can obtain this formula by considering all possibilities of entering failure state E^* before the process $\eta^*(t)$ reaches the critical state. Now it remains only to write the expression for the average time length between two consecutive repairs (renewals). It is equal to

$$\nu_1(h) = \int_0^\infty t \, dQ(t; h), \tag{6.27}$$

where

$$Q(t; h) = Q_{13}(t; h) + Q_{12}(t; h). \tag{6.28}$$

Now we know all the values which appear in the formula for the expected cost per unit time $u(h)$:

$$u(h) = \frac{p_{12}^*(h) + p_{13}^*(h)C}{\nu_1(h)} = \frac{1 - C[1 - p_{13}^*(h)]}{\nu_1(h)}. \tag{6.29}$$

Some numerical results concerning the optimal critical level and minimal cost rate are presented below. They were obtained by use of computerized calculation.

Denote $\min_{h \geq 0} u(h) = u^*(h^*)$, $u_\infty = 1/\nu_1(\infty)$. The ratio

$$\beta = u^*(h^*)/u_\infty \tag{6.30}$$

shows the economy obtained by the optimal use of the prognostic parameter against the costs paid if no PRs are made at all ($h = \infty$). Table 10 gives the values of optimal critical level h^* and the ratio β for different values of C. μ_0 was assumed to be zero. This simplifies the computation because only the ratio $r = \lambda/\mu$ appears in (6.30). The value φ presented in

Table 10

	$r = 8$			$r = 16$			$r = 32$		
C	h^*	β	φ	h^*	β	φ	h^*	β	φ
0.1	1	0.42	0.57	1	0.42	0.56	2	0.46	0.56
0.2	1	0.60	0.77	2	0.64	0.75	3	0.67	0.74
0.3	2	0.76	0.88	3	0.78	0.87	4	0.80	0.86

	$r = 64$			$r = 128$		
C	h^*	β	φ	h^*	β	φ
0.1	3	0.48	0.55	4	0.50	0.55
0.2	5	0.68	0.74	7	0.68	0.74
0.3	7	0.82	0.86	10	0.82	0.85

the table is equal to the ratio $\min_{T>0} d(T)/u_\infty$, where $\min_{T>0} d(T)$ is the minimal expected cost rate for the age replacement based only on the known analytic form of the DF (6.22).

Comparison between β and φ shows that the use of the prognostic parameter gives a considerable economy against the case when only the age is taken into account. So, for $C = 0.2$, $r = 8$, the maintenance based on the prognostic parameter gives saving of 23% in comparison with the optimal age replacement.

We conclude this point with comments on some references. Mercer [62] has investigated a general wear-dependent failure model: shocks occur in accordance with a Poisson process, every shock causes a random amount of damage, the failure rate is proportional to the accumulated damage. This model permits rather simple treatment in the case where every shock causes a constant damage. Some applications of this simplest model to practical situations were given in Section 4 of the book [41]. A preventive maintenance scheme based on the observation of the accumulated wear was proposed in the author's paper (1967).†

Recently there has appeared the work of Taylor [80] which contains a proof that in the case of Mercer's general model, the optimal preventive maintenance rule is: replace either upon the failure or when the damage first exceeds the critical control level.

†Preventive Maintenance Using Prognosis Parameter, Engineering Cybernetics, no. 1, 1967, pp. 54–63.

6.6. *Linear Random Process* $\eta(t) = \mathbf{b}t$ *as a Prognostic Parameter*

Now we consider the situation when the behaviour of the prognostic parameter might be described in terms of a linear-type random process having the form $\eta(t) = \mathbf{b}t$, where $\mathbf{b}$ is a random variable with known DF. During the system's operation it is possible to observe the sample curve of the process $\eta(t)$. This observation gives information about the value $\mathbf{b}$ in the ongoing cycle of operation. According to the definition (6.1) of the prognostic parameter,

$$\underset{\Delta t \to 0}{\boldsymbol{P}}\{t < \boldsymbol{\tau} \leqslant t + \Delta t \,|\, \eta(t) = \eta_0, \boldsymbol{\tau} > t\} = \lambda(\eta_0)\Delta t + o(\Delta t).$$

The condition $\eta(t) = \eta_0$ is equivalent to the choice $\mathbf{b} = \mathbf{b}_0 = \eta_0/t$. Therefore

$$\underset{\Delta t \to 0}{\boldsymbol{P}}\{t < \boldsymbol{\tau} \leqslant t + \Delta t \,|\, \mathbf{b} = b_0, \boldsymbol{\tau} > t\} = \lambda(b_0 t)\Delta t + o(\Delta t). \tag{6.31}$$

In this formula, $\lambda(b_0 t)$ is the system's failure rate when the value b_0 plays a role of a time scale. Observing in a given cycle one sample curve of the prognostic process $\eta(t)$, we obtain the information "how rapidly the lifetime runs" for the system in this cycle. Thus, small values of b_0 means that "time is running slowly" and the system probably will not fail for a long time.

It is well known that the DF of the system's life span may be expressed through its failure rate in the following form:

$$F(t) = 1 - \exp\left[-\int_0^t \lambda(y)\,\mathrm{d}y\right].$$

If it is known from observation that $\mathbf{b} = b_0$, one can state that in a given cycle of system's work it has the following DF of failure-free operation time:

$$F(t) = F_{b_0}(t) = 1 - \exp\left[-\int_0^t \lambda(b_0 y)\,\mathrm{d}y\right]. \tag{6.32}$$

We now turn to the procedure of preventive maintenance. Let us fix a certain critical level h for the prognostic parameter $\eta(t)$ and agree on making a PR when this level is reached, if before this moment the system has not failed. Otherwise, as usual, an ER will be done. The times for execution of PR and ER are respectively, $t_p(h)$ and 1.

Let us obtain the formula for OR. Assume that the value b_0 is known. Then the average length of the active period is

$$T_w(b) = T_b(1 - F_b(T_b)) + \int_0^{T_b} t \, dF_b(t) = \int_0^{T_b} (1 - F_b(t)) \, dt, \tag{6.33}$$

where $T_b = h/b$ – the time for reaching the level h. The average length of the repair time is

$$T_{rep}(b) = (1 - F_b(T_b))t_p(h) + F_b(T_b). \tag{6.34}$$

Now, if **b** is a random variable distributed according to

$$\boldsymbol{P}\{\mathbf{b} = b_l\} = q_l, \qquad l = 1, 2, \ldots, \tag{6.35}$$

then the average values of T_w and T_{rep} will be

$$T_w = \sum_l T_w(b_l)q_l, \qquad T_{rep} = \sum_l T_{rep}(b_l)q_l.$$

From this follows the expression for the OR

$$g(h) = \frac{T_w}{T_w + T_{rep}} = \frac{\sum_l q_l \int_0^{h/b_l} (1 - F_{b_l}(t)) \, dt}{\sum_l q_l \left\{ \int_0^{h/b_l} (1 - F_{b_l}(t)) \, dt + t_p(h) + (1 - t_p(h))F_{b_l}\left(\frac{h}{b_l}\right) \right\}}.$$

The optimal level h^* can be found by means of optimization with respect to h. In the case of degenerate distribution $\boldsymbol{P}\{\mathbf{b} = b^1\} = 1$, we obtain a well-known relation for the age replacement with critical age $T = h/b^1$.

Preventive maintenance which uses the prognostic parameter $\eta(t) = \mathbf{b}t$ is, in fact, a "time-serving policy": when the process $\eta(t)$ is observed even over a very short time, the slope of the line $\eta(t) = \mathbf{b}t$ becomes evident, and, therefore, there is information about the "time scale" in a given cycle of work. The choice of a certain level h means that age replacement is used, in fact, but the critical age is larger for those realizations when the life span of the system is longer.

6.7. *Effectiveness of Preventive Maintenance Using Prognostic Parameter*

The effect of the use of the prognostic parameter depends on how closely the changes of the prognostic parameter and the probability of system's failure are connected. It is easy to imagine a "perfect" prognostic parameter having the following property: for $\eta(t) < \eta_0$, the probability of system's failure is equal to zero; immediately after $\eta(t)$ exceeds the level η_0, the system fails. One may give an example of the opposite nature: in the formula (6.1) $\lambda(\eta)$ does not quite depend on η, i.e. $\lambda(\eta) = \lambda = \text{Const.}$ This is a situation when the life span has an exponential distribution and the prognostic parameter, in fact, does not predict anything.

One practical example is worth mentioning here. For some parts of civil aviation aircraft, the repair rule was based on the accumulated work time, i.e. on the total amount of "time in the air". Later, after certain analysis of failures occurring over a long period of operation, it became obvious that for some parts it was more expedient to base the decision about replacement on the total number of landings. This is related mainly to those parts which failed from fatigue, and the accumulation of fatigue damages was caused mainly by overloading of some parts during landing. In our terms, the number of landings is a better prognostic parameter than the total "time in the air".

Thorough study of the model of failure and discovery of those parameters which are intimately connected with the failure are, generally speaking, the most important factors in preventive maintenance.

7. *The Problem of Finding the Optimal Inspection Period and Optimal "Depth" of Preventive Maintenance*

7.1. *Statement of the Problem*

In this section, we shall study a model of preventive maintenance which is a generalization of the model considered in Section 2 of this chapter.

A system will be considered which behaves according to a Markov process with a finite number of states. The reader is reminded that Section 2 dealt with the particular case of a system having only three states.

An increase in the number of states makes it possible to consider a wider class of repair rules. Thus, in the general case, the repair consists of transferring the system from a "dangerous" state E_i to a safer state $E_{\varphi(i)}$. The repair rule $E_i \to E_{\varphi(i)}$ – the "depth of the repair" – is not fixed beforehand and the function $\varphi(i)$ must be found. Therefore, in that situation not only the inspection times but the "depth of repair" must be optimally chosen. In addition, unlike Section 2, an infinite time horizon will be considered.

Papers [52] and [53] deal with situations similar to those described above. Let us turn to the formal statement of the problem.

The system having $(n+1)$ different states is given. The states are numbered $E_0, E_1, \ldots, E_n^*$. The transitions $E_i \to E_j$ are made according to a certain time-homogeneous Markov process $\eta(t)$ which is defined by means of transition rates $\lambda_{ij}\,(i \neq j)$:

$$\underset{\Delta t \to 0}{P}\,\{\eta(t+\Delta t)=E_j \,|\, \eta(t)=E_i\}$$
$$= \begin{cases} \lambda_{ij}\Delta t + o(\Delta t), & i=0,\ldots,n-1, \quad j\neq i; \\ 1-\sum_{j\neq i}\lambda_{ij}\Delta t + o(\Delta t), & i=0,\ldots,n-1, \quad j=i. \end{cases} \tag{7.1}$$

The state E_n^*, reachable from all other states, represents the failure of the system and has formal properties of an absorbing state: $\lambda_{ni}=0$ for all i. It is assumed that a signal is given immediately after the system enters the failure state. The state E_0 corresponds to an entirely new system.

The repair rule is the following. If at a certain time t_0, the system state E_i is known, then the system is transferred into another state $E_{\varphi(i)}$ and put into operation. The time necessary for the inspection and repair is assumed to be negligible. Then, after completing the repair, a time period $\Delta K_{\varphi(i)}$ is fixed as the time for carrying out the next inspection. That is, the next inspection is planned at time $t_0 + K_{\varphi(i)}\Delta$. Here $K_{\varphi(i)}$ is a certain integer and Δ is a short time interval (e.g. one hour) which will be assumed to be a unit-time interval.

If during the time interval $(t_0,\ t_0+K_{\varphi(i)})$ a signal of failure was not received, the inspection is made at time $t=t_0+K_{\varphi(i)}$; the actual state of the system E_j is discovered and the repair $E_j \to E_{\varphi(j)}$ is performed, after which the next inspection time is fixed, and so on. If at moment $\boldsymbol{\tau}$,

$$t_0 < \boldsymbol{\tau} < t_0 + K_{\varphi(j)}$$

the failure occurs, then the ER will be made immediately. Here the ER consists of transferring the system from the failure state E_n^* to some other state $E_{\varphi(n)}$. After that, a period for the next inspection $K_{\varphi(n)}$ is fixed and the service process goes on.

The rule of performing the repairs and inspections (strategy) can be written in the following general form:

$$\boldsymbol{\sigma} = (\sigma_0, \sigma_1, \ldots, \sigma_n) = (E_{\varphi(i)}, K_{\varphi(i)}; i = 0, 1, \ldots, n). \tag{7.2}$$

Figure 36a shows an example of a system's functioning process. The system has four states E_0, E_1, E_2, E_3^*, and is governed by the strategy

$$\boldsymbol{\sigma} = (E_0, K_0; E_1, K_1; E_0, K_0; E_1, K_1); \qquad K_0 = 4, \quad K_1 = 3. \tag{7.3}$$

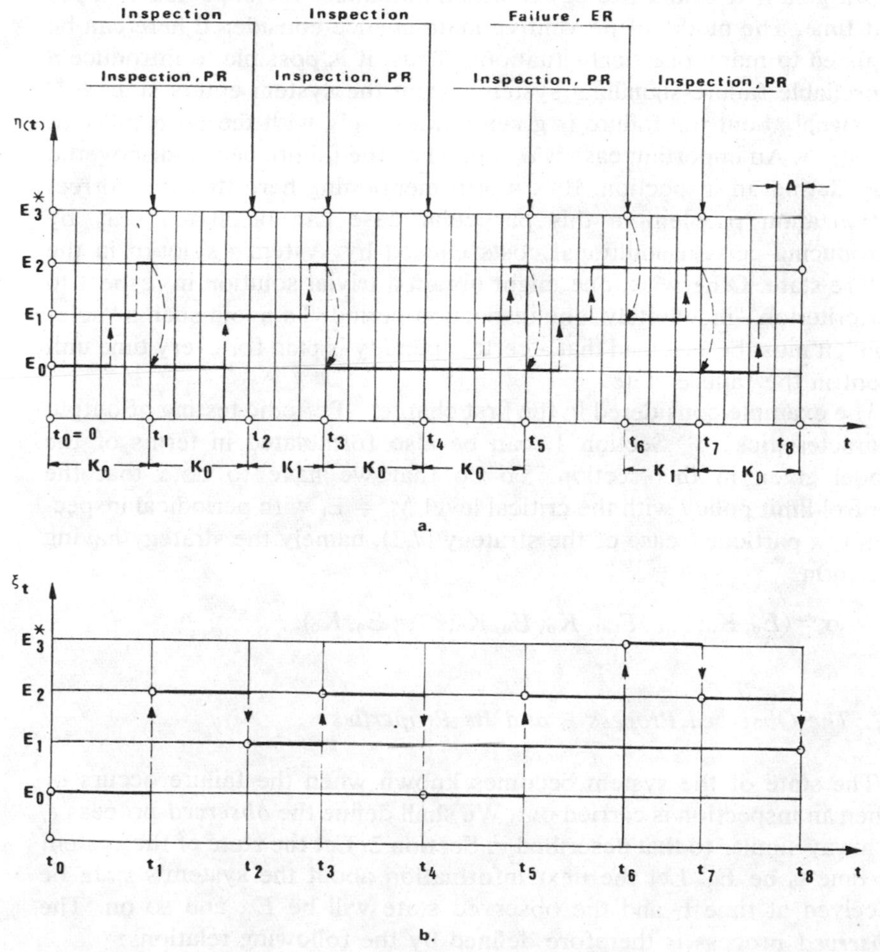

Figure 36. (a) Sample function of the process $\eta(t)$ governed by the strategy (7.3). (b) Sample function of the corresponding observed process ξ_t.

All kinds of repairs and inspections have certain costs. It will be assumed that the cost for ER is c_a, for inspection c_s and the cost for the repair having form $E_i \to E_j$ is equal to c_{ij}.

Thus, for the record plotted in Figure 36a the following total cost will be counted:

$$c_s + c_{20} + c_s + c_s + c_{20} + c_s + c_s + c_{20} + c_a + c_{31} + c_s + c_{20} + c_s = 4c_{20} + c_{31} + 7c_s.$$

Our goal is to find a strategy $\boldsymbol{\sigma}$ which minimizes the expected cost per unit time. The model of preventive maintenance considered here can be adjusted to many practical situations. Thus, it is possible to introduce a nonreliable failure-signalling system: when the system enters state E_n^*, the signal about the failure is given immediately with the probability α, $0 < \alpha < 1$. An important case is $\alpha = 0$, when the failure can be discovered only during an inspection. It is worth mentioning here that the correct optimization problem in this particular case can be stated only by introducing certain additional costs caused by system's sojourn in the failure state. Otherwise, one might obtain a trivial solution in respect to our criterion – indefinitely long inspection period. To avoid such a "solution", it must be assumed that a certain penalty is paid for every time unit spent in the failure state.

The example considered in the first chapter "Periodic testing of output characteristics ..." Section 1, can be also formulated in terms of the model given in this section. To do that we have to note that the control-limit policy with the critical level $M_c = E_l$ with periodical inspection is a particular case of the strategy (7.3), namely the strategy having the form

$$\boldsymbol{\sigma} = (E_0, K_0; \ldots, E_{l-1}, K_0; E_0, K_0; \ldots; E_0, K_0).$$

7.2. *The Observed Process ξ_t and Its Properties*

The state of the system becomes known when the failure occurs or when an inspection is carried out. We shall define the *observed* process ξ_t in a way similar to that described in Section 2. Let the state of the system at time $\mathbf{t}_0$ be E_{i_0}. Let the next information about the system's state be received at time $\mathbf{t}_1$ and the observed state will be E_{i_1}, and so on. The *observed* process is therefore defined by the following relations:

$$\xi_{t_0} = E_{i_0}; \qquad \xi_t = E_{i_0}, \quad \text{for } \mathbf{t}_0 \leq t < \mathbf{t}_1;$$

$$\xi_{t_1} = E_{i_1}; \qquad \xi_t = E_{i_1}, \quad \text{for } \mathbf{t}_1 \leqslant t < \mathbf{t}_2, \ldots$$

It will be assumed that the transitions in the process ξ_t take place at moments $\mathbf{t}_i - 0$, where $\mathbf{t}_i$ are the random time moments when the system is observed. A sample curve of the process ξ_t corresponding to the service process plotted in Figure 35a is shown in Figure 35b.

When the strategy $\boldsymbol{\sigma}$ is fixed (see (7.2)), the process ξ_t is completely defined in the probabilistic sense. (Of course, the process $\eta(t)$ describing the behaviour of the nonrepaired system is assumed to be Markovian.) It is clear that ξ_t is a SMP and $\boldsymbol{\sigma}$ is a strategy of the 1st type.

Let $P_{ij}(t)$ be the transition probability in the process $\eta(t)$:

$$P_{ij}(t) = \boldsymbol{P}\{\eta(t_1 + t) = E_j | \eta(t_1) = E_i\}. \tag{7.4}$$

According to strategy $\boldsymbol{\sigma}$, the system is transferred from state E_i to state $E_{\varphi(i)}$ and the next inspection is planned after time $K_{\varphi(i)}$. During that time, a transition from $E_{\varphi(i)}$ into E_j will occur in the process $\eta(t)$ with the probability $P_{\varphi(i),j}(K_{\varphi(i)})$ and that is exactly the probability that the next observed state will be E_j. Therefore, the matrix of transition probabilities in process ξ_t has the form:

$$\boldsymbol{P}(\boldsymbol{\sigma}) = \|p_{ij}(\sigma_i)\| = \|P_{\varphi(i),j}(K_{\varphi(i)})\|. \tag{7.5}$$

Now let us ascertain what is the DF of the transition time in process ξ_t. Obviously, the transition $E_i \to E_j$, $j \neq n$, has the degenerate DF with a jump at the point $K_{\varphi(i)}$. As for the transition into state E_n^*, we note that the transition probability $P_{\varphi(i),n}(\theta)$ calculated for the process $\eta(t)$ is in respect to process ξ_t the probability of the following event: "a transition from E_i into E_n took place and the transition time was less than θ". Therefore,

$$\boldsymbol{P}\{\boldsymbol{\tau}(i,j) \leqslant t; \sigma_i\} = F_{ij}(t; \sigma_i)$$

$$= \begin{cases} \left.\begin{matrix} 0, & t < K_{\varphi(i)} \\ 1, & t \geqslant K_{\varphi(i)} \end{matrix}\right\} & j \neq n; \\ \left.\begin{matrix} \dfrac{P_{\varphi(i),n}(t)}{P_{\varphi(i),n}(K_{\varphi(i)})}, & t < K_{\varphi(i)} \\ 1, & t \geqslant K_{\varphi(i)} \end{matrix}\right\} & j = n. \end{cases} \tag{7.6}$$

Now we shall calculate $\nu_i(\sigma_i)$, the ME of the duration of one step in the process ξ_t. Using (7.6), (7.5) and formula (3.42) from Chapter 1, we obtain

$$\nu_i(\sigma_i) = \sum_{j=0}^{n} p_{ij}(\sigma_i) \int_0^\infty \tau \, dF_{ij}(\tau; \sigma_i)$$

$$= \sum_{j=0}^{n-1} P_{\varphi(i),j}(K_{\varphi(i)})K_{\varphi(i)} + \int_0^{K_{\varphi(i)}} \tau \, dP_{\varphi(i),n}(\tau)$$

$$= K_{\varphi(i)}(1 - P_{\varphi(i),n}(K_{\varphi(i)})) + \int_0^{K_{\varphi(i)}} \tau \, dP_{\varphi(i),n}(\tau)$$

$$= \int_0^{K_{\varphi(i)}} (1 - P_{\varphi(i),n}(\tau)) \, d\tau. \quad (7.7)$$

It remains only to find the expressions for the costs. The following one-step penalties will be defined for state E_i:

$$\omega_i(\sigma_i) = \begin{cases} c_{i,\varphi(i)} + c_s, & i = 0, 1, \ldots, n-1; \\ c_{n,\varphi(n)} + c_a, & i = n. \end{cases} \quad (7.8)$$

In other words, the one-step cost for state E_i includes costs for the repair and for the inspection (or failure).

Now everything is ready for finding the optimal strategy. If the number K_{max} and the number of system's states are large, then the total number of different stationary strategies is also very large and it is out of the question to use formula (3.42) from Chapter 1 directly. It would be more convenient to use Howard's algorithm (see Chapter 1, Section 3.7). We proceed with an example which illustrates the use of this very efficient algorithm.

7.3. Example

A system having three states E_0, E_1 and E_2^* will be considered. The process $\eta(t)$ will be assumed to be Poissonian, and λ is the transition rate for states E_0 and E_1. The periods between inspections can be Δ or 2Δ. In state E_1 there are two possibilities: to do nothing or to do the repair $E_1 \to E_0$. After a failure, two kinds of ER can be done: $E_2^* \to E_1$ or

$E_2^* \to E_0$. The service strategy has the following general form:

$$\boldsymbol{\sigma} = (\sigma_0, \sigma_1, \sigma_2) = \{E_0, K_0; E_{\varphi(1)}, K_{\varphi(1)}; E_{\varphi(2)}, K_{\varphi(2)}\},$$

where $\varphi(1)$, $\varphi(2)$ are equal to 0 or 1; K_0, $K_{\varphi(i)}$ are 1 or 2. There are two CAs for states E_0 and E_1 and four CAs for E_2^*. Therefore, in this simple example there are $2 \times 4 \times 4 = 32$ different strategies. One may discover at once that among those 32 strategies there are some which are "unpromising". Thus, let us compare, for example, two strategies:

$$\boldsymbol{\sigma}' = \{E_0, \Delta; E_1, 2\Delta; E_0, \Delta\} \quad \text{and} \quad \boldsymbol{\sigma}'' = \{E_0, \Delta; E_1, \Delta; E_0, \Delta\}.$$

It follows from the statement of the problem that the DF of the transition time $E_1 \to E_2^*$ is exponential. Hence, if the system starts to operate with the state E_1, then there is no need to inspect it. (We must also remember that there would be information if the system fails; if not, an inspection would be unnecessary.) But according to the above, an inspection must be done. A better solution would be to postpone the inspection as long as possible, or, in other words, $K = 2$ is always better than $K = 1$, and $\boldsymbol{\sigma}'$ is better than $\boldsymbol{\sigma}''$.

The following costs will be counted: $c_a = 1$, cost for a failure; $c_s = 0.2$, cost for the inspection; $c_{20} = 0.6$ – cost for the repair of type $E_2^* \to E_0$ and $c_{21} = c_{10} = 0.3$, cost for the repair $E_2^* \to E_1$ and $E_1 \to E_0$.

It is easy to check that in our process $\eta(t)$, the transition probabilities are

$$\begin{aligned} &p_{00}(t) = e^{-\lambda t}; \qquad p_{01}(t) = \lambda t\, e^{-\lambda t}; \\ &p_{02}(t) = 1 - e^{-\lambda t}(1 + \lambda t); \qquad\qquad (7.9) \\ &p_{11}(t) = e^{-\lambda t}; \qquad p_{12}(t) - 1 - e^{-\lambda t}. \end{aligned}$$

If we assume $\lambda = 1$, $\Delta = 0.5$, $2\Delta = 1.0$, we obtain the following table containing all needed transition probabilities

$E_i \to E_j$	$p_{ij}(\Delta)$	$p_{ij}(2\Delta)$
$E_0 \to E_0$	0.606	0.368
$E_0 \to E_1$	0.303	0.368
$E_0 \to E_2^*$	0.091	0.264
$E_1 \to E_1$	0.606	0.368
$E_1 \to E_2^*$	0.394	0.632

Assume, for example, the CA in state E_1 is $\sigma_1 = (E_0, 2\Delta)$. That means the PR of type $E_1 \to E_0$ is followed by an inspection done after 2Δ. Therefore, the probability of the transition $E_1 \to E_j$ in the process ξ_t is equal to $P_{0j}(2\Delta)$.

The one-step costs are calculated according to formula (7.8). Thus,

$$\omega_1(\sigma_1) = c_{1,\varphi(1)} + c_s = c_{10} + c_s = 0.5.$$

Finally, we calculate ν_i according to formula (7.9), using (7.7). For example, for the previously mentioned CA σ_1 we have

$$\nu_1(\sigma_1) = \int_0^{2\Delta} (1 - p_{02}(t))\,dt = \int_0^1 e^{-t}(1+t)\,dt = 0.896.$$

All the necessary information is shown in Table 11.

7.3.1. *The first iteration cycle*

1st step. Let us choose the strategy

$$\boldsymbol{\sigma}_1 = (E_0, 0.5; E_1, 0.5; E_1, 0.5). \tag{7.10}$$

Table 11 gives

Table 11

State of ξ_t observed at t_k	CA σ_i for state E_i	Transition probabilities for the process ξ_t $P\{\xi_{t_{k+1}} = E_j \mid \xi_{t_k} = E_i ; \sigma_i\}$			One-step costs $\omega_i(\sigma_i)$	The average transition time $\nu_i(\sigma_i)$ for state E_i
		$p_{i0}(\sigma_i)$	$p_{i1}(\sigma_i)$	$p_{i2}(\sigma_i)$		
E_0	(E_0, Δ)	0.606	0.303	0.091	0.200	0.484
	$(E_0, 2)$	0.368	0.368	0.264	0.200	0.896
E_1	(E_1, Δ)	0	0.606	0.394	0.200	0.394
	$(E_1, 2\Delta)$	0	0.368	0.632	0.200	0.632
	(E_0, Δ)	0.606	0.303	0.091	0.500	0.484
	$(E_0, 2\Delta)$	0.368	0.368	0.264	0.500	0.896
E_2^*	(E_1, Δ)	0	0.606	0.394	1.300	0.394
	$(E_1, 2\Delta)$	0	0.368	0.632	1.300	0.632
	(E_0, Δ)	0.606	0.303	0.091	1.600	0.484
	$(E_0, 2\Delta)$	0.368	0.368	0.264	1.600	0.896

$$\omega_0 = 0.200; \quad \nu_0 = 0.484; \quad p_{00} = 0.606; \quad p_{01} = 0.303;$$
$$p_{02} = 0.091;$$
$$\omega_1 = 0.200; \quad \nu_1 = 0.394; \quad p_{10} = p_{20}; \quad p_{11} = p_{21} = 0.606;$$
$$p_{12} = 0.394; \tag{7.11}$$
$$\omega_2 = 1.300; \quad \nu_2 = 0.394; \quad p_{22} = 0.394.$$

(Indices σ_i denoting the strategy are omitted for the sake of simplicity.)

2nd step. The system of equations (3.48) from Chapter 1 is written down with values ω_i, ν_i, p_{ij} given by (7.11). We put $w_2 = 0$:

$$\begin{aligned} w_0 + 0.484g &= 0.200 + 0.606w_0 + 0.303w_1; \\ w_1 + 0.394g &= 0.200 \qquad\qquad + 0.606w_1; \\ +0.394g &= 1.300 \qquad\qquad + 0.606w_1. \end{aligned} \tag{7.12}$$

The solution is

$$w_0 = -2.312; \quad w_1 = -1.100; \quad g = 1.607.$$

3rd step. For every state E_i, the CA σ_i which minimizes the expression of type (3.49), Chapter 1, is found. For E_0:

$$z(\sigma_0) = \frac{\omega_0 + p_{00}w_0 + p_{01}w_1 - w_0}{\nu_0} = \begin{cases} 1.607 & \text{for} \quad \sigma_0 = (E_0, \Delta); \\ 1.402 & \text{for} \quad \sigma_0 = (E_0, 2\Delta). \leftarrow \end{cases}$$

For E_1:

$$z_1(\sigma_1) = \frac{\omega_1 + p_{10}w_0 + p_{11}w_1 - w_1}{\nu_1} = \begin{cases} 1.607 & \text{for} \quad \sigma_1 = (E_1, \Delta); \\ 1.416 & \text{for} \quad \sigma_1 = (E_1, 2\Delta); \\ -0.278 & \text{for} \quad \sigma_1 = (E_0, \Delta); \leftarrow \\ 0.384 & \text{for} \quad \sigma_1 = (E_0, 2\Delta). \end{cases}$$

For E_2:

$$z(\sigma_2) = \frac{\omega_2 + p_{20}w_0 + p_{21}w_1}{\nu_2} = \begin{cases} 1.607 & \text{for} \quad \sigma_2 = (E_1, \Delta); \\ 1.416 & \text{for} \quad \sigma_2 = (E_1, 2\Delta); \\ -0.278 & \text{for} \quad \sigma_2 = (E_0, \Delta); \leftarrow \\ 0.384 & \text{for} \quad \sigma_2 = (E_0, 2\Delta). \end{cases}$$

The arrows indicate the CAs which provide the minimal values of the ratios $z(\sigma_i)$. Therefore, the first improvement of strategy $\boldsymbol{\sigma}_1$ is

$$\boldsymbol{\sigma}_2 = (E_0, 2\Delta\,;\, E_0, \Delta\,;\, E_0, \Delta). \tag{7.13}$$

7.3.2. The second iteration cycle

1st step. For the strategy $\boldsymbol{\sigma}_2$ we find from Table 11

$$\begin{aligned} &\omega_0 = 0.200; \quad \nu_0 = 0.896; \quad p_{00} = 0.368; \quad p_{01} = 0.368;\\ &\omega_1 = 0.500; \quad \nu_1 = 0.484; \quad p_{10} = p_{20} = 0.606;\\ &p_{11} = p_{21} = 0.303;\\ &\omega_2 = 1.600; \quad \nu_2 = 0.484. \end{aligned} \tag{7.14}$$

2nd step. The system (3.48), Chapter 1, is the following:

$$\begin{aligned} w_0 + 0.896g &= 0.200 + 0.368w_0 + 0.368w_1,\\ w_1 + 0.484g &= 0.500 + 0.606w_0 + 0.303w_1,\\ 0.484g &= 1.600 + 0.606w_0 + 0.303w_1. \end{aligned} \tag{7.15}$$

Its solution is

$$w_0 = -1.452; \qquad w_1 = -1.100; \qquad g = 0.799.$$

3rd step. The CAs which minimize ratios $z(\sigma_i)$ are:

For E_0:

$$z(\sigma_0) = \begin{cases} 0.905 & \text{for} \quad \sigma_0 = (E_0, \Delta);\\ 0.796 & \text{for} \quad \sigma_0 = (E_0, 2\Delta). \leftarrow \end{cases}$$

For E_1:

$$z(\sigma_1) = \begin{cases} 1.607 & \text{for} \quad \sigma_1 = (E_1, \Delta);\\ 1.416 & \text{for} \quad \sigma_1 = (E_1, 2\Delta);\\ 0.799 & \text{for} \quad \sigma_1 = (E_0, \Delta);\\ 0.737 & \text{for} \quad \sigma_1 = (E_0, 2\Delta). \leftarrow \end{cases}$$

For E_2^* $z(\sigma_1) \equiv z(\sigma_2)$. The second improvement of the strategy is

$$\boldsymbol{\sigma}_3 = (E_0, 2\Delta\,;\, E_0, 2\Delta\,;\, E_0, 2\Delta). \tag{7.16}$$

7.3.3. The third iteration cycle

1st step. For strategy $\boldsymbol{\sigma}_3$ we get from Table 11

$\omega_0 = 0.200; \quad \nu_0 = \nu_1 = \nu_2 = 0.896; \quad p_{00} = p_{01} = p_{20} = p_{10} = p_{11} = p_{21} = 0.368; \quad \omega_1 = 0.500; \quad \omega_2 = 1.600.$

2nd step. The system (3.48), Chapter 1 takes the form:

$$\begin{aligned} w_0 + 0.896g &= 0.200 + 0.368w_0 + 0.368w_1, \\ w_1 + 0.896g &= 0.500 + 0.368w_0 + 0.368w_1, \\ 0.896g &= 1.600 + 0.368w_0 + 0.368w_1. \end{aligned} \tag{7.17}$$

Its solution is

$$w_0 = -1.400; \qquad w_1 = -1.100; \qquad g = 0.758.$$

3rd step. We make sure that ratios $z(\sigma_i)$ are minimized by the same CA as in the previous iteration cycle. Hence, (7.16) is the optimal strategy and the minimal average cost rate per unit time is $\min_\sigma g = 0.758$.

It is interesting to find out what would be the average cost rate if we distinguish only two states, E_0 and E_2^*, in our system and preserve the same stochastic properties, penalties, etc. We are forced to do a PR at every inspection, because there is no exact information about the system's state (E_0 or E_1). When the inspection period is T we obtain the following expression for the expected cost per unit time:

$$g(T) = \frac{(1 - p_{02}(T))a_2 + p_{02}(T)a_1}{\int_0^T (1 - p_{02}(t))\,dt}. \tag{7.18}$$

Here T is equal to 0.5 or 1.0, $a_2 = c_{10} + c_s = 0.5$; $a_1 = c_0 + c_{20} = 1.6$. Substituting (7.9) into (7.18) we obtain

$$g(0.5) = 1.24; \qquad g(1) = 0.88.$$

From this follows the optimal $T = 1$. In this case, the cost rate is 15% greater than for the optimal strategy.

7.4. Concluding Remarks and Possible Generalizations

Remark 1. Considering the class of available strategies, we have restricted ourselves to the condition that the inspection period after finding the system in the state E_i depends only on the state after repair $E_{\varphi(i)}$. Let us assume that we have at our disposal a broader class of strategies, having the form $\{\varphi(i), T_{i,\varphi(i)}; i = 1, n\}$. In other words it is assumed that the inspection period depends both on the observed state E_i and the state into which the system was transferred, $E_{\varphi(i)}$. Does that extension permit a decrease in the average cost per unit time?

Assume that for two states, E_i and E_j, the optimal strategy prescribes the same kind of repair $\varphi(i) = \varphi(j)$, but different inspection periods T_i and T_j, $T_i \neq T_j$. If that happens, it is worth considering the extension, although, by using Howard's relations for finding the optimal strategy, one may prove that the answer to the question is negative. The proof of this is given in Problem 6 of Section 10.

The reader may find this to be absolutely clear on the basis of intuitive reasoning.

Remark 2. However effective Howard's algorithm may be, the following question remains valid: is it possible to restrict ourselves, on the basis of some theoretical considerations, to a simpler class of strategies? If the answer is positive, we can essentially shorten the calculation procedure.

In the practice of preventive maintenance, the so-called threshold or control-limit policy is often used. For this policy, the function $\varphi(i)$ has the following simple form:

$$\varphi(i) = i \quad \text{for} \quad i \leq i^*, \qquad \varphi(i) = 0 \quad \text{for} \quad i > i^*.$$

The intuitive reason for using this is clear: the total repair has to be done if and only if the deterioration of the system exceeds some critical level E_i^*. If only periodic inspections are made, then the number of different stationary strategies is relatively very small, because the strategy is entirely defined by the choice of the critical state.

There are several works devoted to the question of the optimality of control-limit strategies among all stationary strategies. Thus, [10, Chapter 5, Section 4], [55] and [29, Chapter 9] consider the question in respect to controlled Markov chains. References [40] and [52] contain some facts about optimality conditions for SMP. Reference [22] reveals the optimality of a threshold policy in a special controlled queueing system having a nonreliable service device. We think that this point is of major practical and theoretical interest and are devoting a special discussion in this chapter to this question.

Remark 3. Let us mention two directions in which the problem treated in this section might be generalized.

First, it is easy to imagine a situation where an extra cost d_i must be paid for a unit of time which the system has spent in state E_i (see also a similar discussion at the end of Section 3). In Problem 4 of Section 10 such a situation will be considered and the corresponding solution presented.

Secondly, let us assume that the system has several "undesirable" states and it is necessary to restrict in some way the time the system stays in them. Practically speaking, these states might correspond to those in which it is difficult to carry out repairs or other kinds of services. Formally, the presence of such states means that the problem of finding the optimal strategy has to be solved in the presence of one or several constraints. It turns out that in this case Howard's algorithm is not applicable and another mathematical approach must be given. We will discuss this in the next section.

8. Optimal Control of a Semi-Markov Process in the Presence of Constraints on State Probability

8.1. Statement of the Problem

Let us assume that one or more states of a system have properties that require restricting the time in which the system stays in those states. Here we shall deal only with the formal concepts of this peculiarity and not the practical aspects of the situation. The model of controlled SMP in the presence of several linear constraints has the following general form:

$$\sum_{j=1}^{n} P_j C_j^{(S)} \leq C_S, \quad S = \overline{1, S_0}, \tag{8.1}$$

P_j are stationary probabilities for the SMP.†

The particular case of constraints is simply a restriction of the form

$$P_r < P_{\max} \tag{8.2}$$

related only to one "undesirable" state.

Our aim is to find a stationary Markovian strategy $\boldsymbol{\sigma} = (\sigma_1, \ldots, \sigma_n)$ which provides the minimal average cost per unit time, while preserving the given constraints of form (8.1).

Up to now, we have considered only strategies prescribing the execu-

†P_j has the following formal meaning: $P_j = \lim_{t\to\infty} P\{\xi_t = E_j | \xi_0 = E_i\} = \pi_j \nu_j / \Sigma_{i=1}^{n} \pi_i \nu_i$ (see Appendix, A.9).

tion of some definite CA which depends on the observed state, and possibly, the time and the number of the step. Here we need to introduce the *randomized* Markovian stationary strategies. A randomized strategy prescribes the choice of the CA in the state E_i by help of some stochastic mechanism. Formally speaking, the choice of CA is made according to some probability distribution given for the set S_i of all CAs available in state E_i.

In solving all the previous problems we did very well with nonrandomized strategies. There is a formal proof of the fact that in finding the optimal stationary Markovian strategy, without any additional constraints, it is sufficient to look for an optimal policy in the subclass of all nonrandomized policies (see, for example, [29]). Considering the case with constraints, the situation changes. It may happen that the best strategy does not satisfy the restrictions and it becomes necessary to "mix" it with another nonoptimal strategy in order to satisfy the constraints.

The presence of constraints and the necessity to consider a wider class of strategies does not permit the solving of the problem within the framework of the Howard iterative procedure. However, methods of linear programming are found to be useful in this case.

Many works have been published devoted to finding the optimal Markovian strategy of controlled Markov chains and SMP with the help of linear programming (LP). We shall touch only briefly on the results. Papers [47], [60] and [84] assert that the problem of finding an optimal strategy might be reduced to a LP problem. It has also been stated that the optimal strategy belongs to the subclass of nonrandomized strategies. A more detailed study of the LP problem applied to controlled Markov processes has shown that the dual LP problem is closely connected with Howard's algorithm ([29] and [47]).

Throughout our exposition, we have considered only those cases when the controlled SMP has only one closed set of states. If a strategy governing the SMP generates more than one closed subset of states, then the subject becomes essentially more complicated. The problem of finding the optimal stationary strategy for a SMP in a multichain case, with the help of LP, was studied in [24], [29], [73], [74] and [67]. The last reference gives an exhaustive study of this problem.

The problem with constraints was considered in [17] and [27]. Constraints on the frequency of visiting a given state were introduced in these articles. Reference [81] deals with some specific kinds of linear constraints. Our exposition here follows the author's paper [39] in which the case of one ergodic class was investigated.

We will show how the general problem without restrictions can be formulated in terms of LP. Later, the case with constraints having type (8.2) will be considered. Here an interesting property of the optimal strategy can be stated: if there is only one linear constraint, then the optimal strategy remains nonrandomized for all states except for one state; in that state the randomization is made by mixing together two CAs. Finally, the connection between the LP formulation and Howard's algorithm will be briefly investigated.

8.2. Notation

It will be assumed that in every state there are available CAs numbered by 1, 2, ..., K. Instcad of $p_{ij}(\sigma_k)$, we shall use the notation $p_{ij}(k)$; in a similar manner, we write the one-step costs as $\omega_i(k)$ and average transition times as $\nu_i(k)$, $k - 1, \ldots, K$.

Let us introduce the following collection of nonnegative numbers:

$$\bar{\alpha} = \|\alpha_{ij}\|, \qquad i = \overline{1, n}, \qquad j = \overline{1, K},$$
$$\alpha_{ij} \geqslant 0, \qquad \sum_{j=1}^{K} \alpha_{ij} = 1. \tag{8.3}$$

We shall interpret α_{ik}, $k = \overline{1, K}$, as probabilities of selecting the CA with number k in state E_i. The process ξ_t governed by strategy $\bar{\alpha}$ will be, of course, a SMP. All its characteristics will be denoted with symbols having a dash above: $\bar{p}_{ij}$, $\bar{\omega}_i$, $\bar{\nu}_i$.

It is clear that

$$\bar{p}_{ij} = \sum_{k=1}^{K} \alpha_{ik} p_{ij}(k), \qquad \bar{\nu}_i = \sum_{k=1}^{K} \alpha_{ik} \nu_i(k), \qquad \bar{\omega}_i = \sum_{k=1}^{K} \alpha_{ik} \omega_k. \tag{8.4}$$

The expected cost per unit time in that process is equal to

$$\bar{g}(\bar{\alpha}) = \sum_{i=1}^{n} \bar{\pi}_i \bar{\omega}_i \Big/ \sum_{i=1}^{n} \bar{\pi}_i \bar{\nu}_i, \tag{8.5}$$

where $\{\bar{\pi}_i\}$, $i = \overline{1, n}$ is the solution of the following system:

$$\{\bar{\pi}_1, \ldots, \bar{\pi}_n\} = \{\bar{\pi}_1, \ldots, \bar{\pi}_n\} \bar{\mathbf{P}};$$
$$\sum_{i=1}^{n} \bar{\pi}_i = 1, \qquad \bar{\mathbf{P}} = \|\bar{p}_{ij}\|, \quad i, j = \overline{1, n}. \tag{8.6}$$

In the sequel, the following notations will be used:

$$\Omega = (\omega_1(1), \ldots, \omega_1(K); \ldots; \omega_n(1), \ldots, \omega_n(K));$$

$$\Omega_1 = (\Omega; 0);\ \bar{\Omega} = (\bar{\omega}_1/\bar{\nu}_1, \ldots, \bar{\omega}_n/\bar{\nu}_n); \tag{8.7}$$

$$\bar{x} = (x_{11}, \ldots, x_{1K}; \ldots; x_{n1}, \ldots, x_{nK});$$

$$\bar{x}_1 = (\bar{x}, y), \tag{8.8}$$

$$t = (\underbrace{0, \ldots, 0}_{n}, 1), \quad t_1 = (\underbrace{0, \ldots, 0}_{n}, 1, C). \tag{8.9}$$

$$\mathbf{B} = \left\{ \begin{array}{l} 1 - p_{11}(1), \ldots, 1 - p_{11}(K);\ -p_{21}(1), \ldots, -p_{21}(K); \ldots; \\ \qquad\qquad\qquad\qquad -p_{n1}(1), \ldots, -p_{n1}(K) \\ -p_{12}(1), \ldots, -p_{12}(K);\ 1 - p_{22}(1), \ldots, 1 - p_{22}(K); \\ \qquad\qquad\qquad\qquad -p_{n2}(1), \ldots, -p_{n2}(K) \\ \cdot \quad \cdot \quad \cdot \quad \cdot \quad \cdot \quad \cdot \quad \cdot \quad \cdot \quad \cdot \quad \cdot \\ \cdot \quad \cdot \quad \cdot \quad \cdot \quad \cdot \quad \cdot \quad \cdot \quad \cdot \quad \cdot \quad \cdot \\ \cdot \quad \cdot \quad \cdot \quad \cdot \quad \cdot \quad \cdot \quad \cdot \quad \cdot \quad \cdot \quad \cdot \\ -p_{1n}(1), \ldots, -p_{1n}(K); \ldots\ldots; 1 - p_{nn}(1), \ldots, 1 - p_{nn}(K) \\ \nu_1(1), \ldots, \nu_1(K); \ldots\ldots\ldots\ldots; \nu_n(1), \ldots, \nu_n(K) \end{array} \right\} \tag{8.10}$$

$$\mathbf{B}_1 = \left\{ \begin{array}{c|c} \underbrace{\mathbf{B}}_{nK} \ \ (n+1 \text{ rows}) & \begin{array}{c} 0 \\ 0 \\ \vdots \\ 0 \end{array} \\ \hline \varphi_{11}, \ldots, \varphi_{1K}; \ldots; \varphi_{n1}, \ldots, \varphi_{nK} & 1 \end{array} \right\} \tag{8.10*}$$

8.3. *Finding the Optimal Strategy by Means of LP – The Problem without Constraints*

If the SMP has only one ergodic class of states, the following lemma is valid.

Lemma. *Let* $\mathbf{P} = \|p_{ij}\|$, $i, j = \overline{1, n}$, *be the matrix of the imbedded Markov chain,* $\boldsymbol{\pi} = (\pi_1, \ldots, \pi_n)$ *be a nonzero solution of the system*

$$\boldsymbol{\pi} = \boldsymbol{\pi}\mathbf{P}. \tag{8.11}$$

Then the system of equations

$$\left(\frac{\beta_1}{\nu_1}, \ldots, \frac{\beta_n}{\nu_n}\right) = \left(\frac{\beta_1}{\nu_1}, \ldots, \frac{\beta_n}{\nu_n}\right)\mathbf{P}, \tag{8.12}$$

$$\sum_{i=1}^{n} \beta_i = 1.$$

has a unique solution

$$\beta_i = \pi_i \nu_i \Big/ \sum_{i=1}^{n} \pi_i \nu_i. \tag{8.13}$$

According to this lemma, the average cost per unit time in the process ξ_t, which is governed by a randomized strategy, is equal to

$$\bar{g}(\bar{\alpha}) = \bar{\boldsymbol{\beta}}\bar{\Omega}^T, \tag{8.14}$$

where the vector $\bar{\boldsymbol{\beta}}$ satisfies the following system of equations:

$$(\bar{\beta}_1/\bar{\nu}_1, \ldots, \bar{\beta}_n/\bar{\nu}_n) = (\bar{\beta}_1/\bar{\nu}_1, \ldots, \bar{\beta}_n/\bar{\nu}_n)\bar{\mathbf{P}},$$
$$\sum_{i=1}^{n} \bar{\beta}_i = 1. \tag{8.15}$$

(The symbol T denotes transposition.) Let us substitute the values p_{ij} given by (8.4) into (8.15); then we obtain:

$$\bar{\beta}_i/\bar{\nu}_i - \sum_{j=1}^{n}\sum_{r=1}^{K} (\bar{\beta}_j/\bar{\nu}_j)\bar{p}_{ji}(r)\alpha_{jr} = 0, \qquad i = \overline{1, n}. \tag{8.16}$$

Equations (8.16) and (8.13) may be considered as a system of equations in respect to the unknown quantities $\{\alpha_{ij}, \bar{\beta}_i\}$, $j = \overline{1, K}$; $i = \overline{1, n}$.

We are interested in finding a solution which will provide the minimal value for the quantity $\bar{g}(\bar{\alpha})$ given by (8.14). This formulation is not convenient for the solution because the system is not linear in respect to the values α_{ij}. We shall obtain a more convenient situation after introducing new variables

$$x_{jr} = (\bar{\beta}_j/\bar{\nu}_j)\alpha_{jr}; \qquad j = \overline{1, n}, \quad r = \overline{1, K}. \tag{8.17}$$

The condition (8.3) means that

$$\bar{\beta}_j/\bar{\nu}_j = \sum_{r=1}^{K} x_{jr}; \qquad j = \overline{1, n}, \quad x_{jr} \geqslant 0. \tag{8.18}$$

The first n equations in (8.15) will take the form:

$$i = \overline{1, n}: \sum_{r=1}^{K} x_{ir} - \sum_{j=1}^{n} \sum_{r=1}^{K} p_{ji}(r) x_{jr} = 0, \tag{8.19}$$

and the last one because of (8.17) will be as follows:

$$\sum_{i=1}^{n} \sum_{r=1}^{K} \nu_i(r) x_{ir} = \sum_{i=1}^{n} \bar{\beta}_i = 1.$$

Therefore, the system of all constraints containing new quantities will have the form:

$$\begin{aligned} &\mathbf{B}\bar{x}^T = t^T, \\ &\bar{x} \geqslant 0. \end{aligned} \tag{8.20}$$

The cost rate $\bar{g}$ can be expressed in the following manner:

$$\bar{g} = \sum_{i=1}^{n} \bar{\pi}_i \bar{\omega}_i \Big/ \sum_{i=1}^{n} \bar{\pi}_i \bar{\nu}_i = \sum_{j=1}^{n} \frac{\bar{\beta}_j}{\bar{\nu}_j} \bar{\omega}_j = \sum_{r=1}^{K} \sum_{j=1}^{n} \frac{\bar{\beta}_j}{\bar{\nu}_j} \alpha_{jr} \omega_j(r)$$

$$= \sum_{r=1}^{K} \sum_{j=1}^{n} x_{jr} \omega_j(r) = \bar{x} \Omega^T. \tag{8.21}$$

The lemma asserts that $\Sigma_{r=1}^{K} x_{jr} > 0$ for all j. This makes it possible, after finding a solution $\bar{x}$ satisfying (8.20), to return to the initial variables

$$\alpha_{jr} = x_{jr} \Big/ \sum_{r=1}^{K} x_{jr}. \tag{8.22}$$

Hence, the problem of finding an optimal randomized strategy is reduced to the following LP problem: to find a vector $\bar{x}$ which minimizes (8.21) in the presence of constraints (8.20). Theorem 1 asserts the main property of the optimal strategy.

Theorem 1. *There is a solution of the following problem:*

$$\min \bar{x}\Omega^T$$

$$\mathbf{B}\bar{x}^T = t^T,$$

$$\bar{x} \geqslant 0,$$

having the property that for every i there is one and only one value $x_{ik} > 0$ (other x_{ik}, $k = 1, \ldots, n$, are equal to zero).

The proof of this theorem is omitted because its main points are contained in Theorem 2. The reader is referred to the paper of Wolfe and Dantzig [84] for this proof.

Theorem 1 asserts, therefore, that there is an optimal strategy which is in fact, nonrandomized: for every state only one CA is selected with the probability 1. This becomes obvious when we go back from variables x_{ik} to α_{ik} (see (8.22)).

8.4. The Problem with a Constraint

Assume that we have to find $\bar{\alpha}$ which minimizes $\bar{g}(\bar{\alpha})$ by the presence of the following additional constraint:

$$\sum_{j=1}^{n} \bar{P}_j c_j = \sum_{j=1}^{n} \frac{\bar{\nu}_j \bar{\pi}_j}{\sum_{1}^{n} \bar{\pi}_i \bar{\nu}_i} c_j < C. \tag{8.23}$$

Denote

$$\varphi_{ik} = c_i \nu_i(k). \tag{8.24}$$

Theorem 2. *Let $\bar{x}_1$ be a vector which minimizes*

$$\bar{g} = \bar{x}_1 \Omega_1^T \tag{8.25}$$

in the presence of the following constraints:

$$\begin{aligned} \mathbf{B}_1 \bar{x}_1^T &= t_1^T, \\ \bar{x}_1 &\geq 0. \end{aligned} \tag{8.26}$$

Then the vector $\bar{x}_1$ determines a unique strategy $\bar{\alpha}$ providing the minimal average cost rate by satisfying the constraint (8.23).

Proof. It is sufficient to show that the last, $(n+2)$th, equation in (8.26) is equivalent to (8.23). But it is true because

$$\sum_{i=1}^{n} \sum_{r=1}^{K} \varphi_{ir} x_{ir} = \sum_{i=1}^{n} \sum_{r=1}^{K} c_i \nu_i(r) \frac{\bar{\beta}_i}{\bar{\nu}_i} \alpha_{ir} = \sum_{i=1}^{n} c_i \bar{\beta}_i$$

and

$$y \geq 0. \quad \#$$

The next theorem describes the structure of an optimal strategy. Its proof follows the main lines of the proof of the lemma in the cited paper [84].

Theorem 3. *Among the solutions of the problem*

$$\min \bar{g} = \min \{\bar{x}_1 \Omega_1^T\}$$
$$\mathbf{B}_1 \bar{x}_1^T = t_1^T,$$
$$\bar{x}_1 \geq 0,$$

there is a solution $\bar{x}_1$ *having the following property: vector* $\bar{x}_1$ *contains not more than* $(n+1)$ *nonzero elements; in every group of variables* $(x_{i1}, \ldots, x_{iK})$, $i = \overline{1, n}$, *there is at least one nonzero element.*

Proof. Let us consider a submatrix $\mathbf{B}_0$ of matrix $\mathbf{B}_1$ consisting of all columns which correspond to nonzero elements of the vector $\bar{x}_1$ which is a solution of the problem. There is a well-known property of LP problems which states that we can find a solution $\bar{x}_1$ with, say, r nonzero elements, such that the corresponding matrix $\mathbf{B}_0$ will have the rank equal to r. When we consider this solution, we note that the sum of the first n rows in the matrix $\mathbf{B}_0$ is equal to zero. Therefore, the rank of $\mathbf{B}_0$ cannot exceed $(n+1)$ and $r \leq n+1$. Let s rows in $\mathbf{B}_0$ contain elements of type $1 - p_{ii}(k)$. Except for the last two rows, in $\mathbf{B}_0$ there are $(n-s)$ rows not containing elements of type $1 - p_{ii}(k)$; these rows have only nonpositive elements, $-p_{ij}(k)$. From (8.26) it follows that the product of these rows with the positive elements of the vector $\bar{x}_1$ is equal to zero. That means that these $(n-s)$ rows contain only zero elements. The rank of $\mathbf{B}_0$ cannot exceed $(s+1)$, i.e. $r \leq s+1$. At the same time, the number s is equal to the number of groups of variables $\{x_{ij}\}$, $j = \overline{1, K}$, having nonzero elements. From this follows the inequality $s \leq r \leq s+1$ and the fact that each group of variables $\{x_{ir}, r = \overline{1, K}\}$ cannot have more than two positive elements, and that only one group with two positive elements may exist.

Now we shall prove that $r \geq n$. Let r_0 be the number of groups having nonzero elements. Without loss of generality it may be assumed that

$$x_{11} \geq 0,\ x_{12} \geq 0,\ x_{21} > 0, \ldots, x_{r_0 1} > 0 \quad (\text{here } x_{11} + x_{12} > 0).$$

Assume that there exists a solution

$$\mathbf{x}_0 = (x_{11}, x_{12}, x_{21}, x_{31}, \ldots, x_{r_0 1}, z_{r_0+1}, \ldots, z_n),$$

having all $z_m = 0$, $m = \overline{r+1, n}$. After simple transformations, we can rewrite the first equation of (8.26) in the form

$$\tilde{x}_{11} = \tilde{x}_{11}\tilde{p}_{11}(1) + x_{21}p_{21}(1) + \cdots + x_{r_0 1}p_{r_0 1}(1)$$
$$+ z_{r_0+1}p_{r_0+1,1}(1) + \cdots + z_n p_{n1}(1),$$

where

$$\tilde{x}_{11} = (x_{11} + x_{12}), \tilde{p}_{11} = \frac{p_{11}(1)x_{11}}{x_{11} + x_{12}} + \frac{p_{11}(2)x_{12}}{x_{11} + x_{12}}.$$

The same can be done with the first n equations of the system. As a result we shall obtain the following system of equations:

$$(\tilde{x}_{11}, x_{21}, \ldots, x_{r1}, z_{r_0+1}, \ldots, z_n)$$

$$= (\tilde{x}_{11}, x_{21}, \ldots, x_{r1}, z_{r_0+1}, \ldots, z_n) \begin{bmatrix} \tilde{p}_{11}(1), \ldots, \tilde{p}_{1n}(1) \\ p_{21}(1), \ldots, p_{2n}(1) \\ p_{n1}(1), \ldots, p_{nn}(1) \end{bmatrix},$$

where

$$\tilde{p}_{1j} = \frac{p_{1j}(1)x_{11} + p_{1j}(2)x_{12}}{x_{11} + x_{12}}, \qquad j = \overline{1, n}.$$

It is obvious that the matrix of that system is stochastic and that the Markov chain defined by it is an ergodic one. This excludes the possibility for z_j to be equal to zero. Therefore, $r_0 \geq n$, and this completes the proof. #

If it turns out that the optimal vector $\bar{x}_1$ contains $(n + 1)$ nonzero elements and $y = 0$, then the optimal strategy will be randomized. Theorem 3 asserts that randomization is done only in one state by mixing two CAs.

8.5. Several Constraints

When there are several constraints having the form (8.1), the solution of the problem remains in principle the same. An additional row in the matrix **B**

$$\{\varphi_{11}^{(S)}, \ldots, \varphi_{nn}^{(S)}, \overbrace{0, \ldots, 0}^{S-1}, 1, 0, \ldots, 0\}$$

will correspond to the Sth restriction, $S = \overline{1, S_0}$, and the vector $\bar{x}_1$ will contain coordinates $y_1, y_2, \ldots, y_{S_0}$; t_1 will have the form $t_1 = (0, \ldots, 0, 1; C_1, \ldots, C_{S_0})$. By repeating the arguments of Theorem 3, we show that there exists an optimal vector $\bar{x}_1$ having not more than $n + S_0$ positive coordinates, and that every group of variables $\{x_{jr}\}$, $r = \overline{1, K}$, which have a fixed first index, must have at least one positive value.

It may happen that the following constraint must be kept:

$$\sum_{i=1}^{n} \frac{a_i}{\bar{\mu}_{ii}} < C,$$

where $\bar{\mu}_{ii}$ is the average return time into state E_i. The meaning of such a restriction is obvious: the frequency of visiting one or several states is restricted. This case does not cause any difficulties because of the simple relationship between the return time $\bar{\mu}_{ii}$ and the stationary probability $\bar{\pi}_i$ in a SMP:

$$\left[\bar{\pi}_i \Big/ \sum_{i=1}^{n} \bar{\pi}_i \bar{\nu}_i\right]^{-1} = \bar{\mu}_{ii}$$

(See Appendix 2, Theorem 2.) Quantities φ_{ik} in (8.24) must be redefined as $\varphi_{ik} = a_i$, $k = \overline{1, K}$.

In order to demonstrate how the presence of a constraint leads to a randomized strategy, we offer a simple example.

Example. A SMP with two states is given. In state E_1 there are two CAs numbered 1 and 2; in state E_2 there are no alternatives and the only CA has the number 1.

The information concerning the SMP is as follows:

$$\begin{aligned} &E_1: p_{11}(1) = 0, \quad p_{11}(2) = 0; \qquad && E_2: p_{21}(1) = 1, \quad p_{22}(1) = 0.\\ &\quad\; p_{12}(1) = 1, \quad p_{12}(2) = 1; && \\ &\quad\; \nu_1(1) = 1, \quad \nu_1(2) = 3; && \nu_2(1) = 2,\\ &\quad\; \omega_1(1) = -2, \quad \omega_1(2) = -2; && \omega_2(1) = -3. \end{aligned}$$

The constraint has the form:

$$P_2 \leqslant 0.5,$$

or

$$\frac{\bar{\pi}_2 \bar{\nu}_2}{\bar{\pi}_1 \bar{\nu}_1 + \bar{\pi}_2 \bar{\nu}_2} \leqslant 0.5.$$

In this example p_{ij} does not depend on the CA and therefore it is easy to find that $\bar{\pi}_1 = \bar{\pi}_2 = \frac{1}{2}$. It follows that

$$\frac{\bar{\nu}_2}{\bar{\nu}_1 + \bar{\nu}_2} \leqslant 0.5.$$

According to (8.24) we have to set

$$\varphi_{11}=0, \quad \varphi_{12}=0, \quad \varphi_{21}=c_2\nu_2(1)=2.$$

The system (8.26) has the form

$$\begin{Vmatrix} 1 & 1 & -1 & 0 \\ -1 & -1 & 1 & 0 \\ 1 & 3 & 2 & 0 \\ 0 & 0 & 2 & 1 \end{Vmatrix} \begin{Vmatrix} x_{11} \\ x_{12} \\ x_{21} \\ y \end{Vmatrix} = \begin{Vmatrix} 0 \\ 0 \\ 1 \\ 0.5 \end{Vmatrix}$$

$$x_{11}\geqslant 0, \quad x_{12}\geqslant 0, \quad x_{21}\geqslant 0, \quad y\geqslant 0$$

Its solution is:

$$x_{11}=\frac{0.5-5y}{4}, \qquad x_{12}=\frac{0.5+3y}{4}, \qquad x_{21}=\frac{0.5-y}{2}.$$

The condition $x_{ij}\geqslant 0$ leads to the inequality $0\geqslant y\geqslant 0.1$. According to (8.25)

$$\bar{g}=-\left[\frac{0.5-5y}{4}\cdot 2+\frac{0.5+3y}{4}2+\frac{0.5-y}{2}\cdot 3\right]=\frac{-5+10y}{4}$$

and

$$\min_{0\leqslant y\leqslant 0.1}\ \bar{g}=\bar{g}|_{y=0}=-1.25.$$

Thus, $x_{11}=x_{12}$; $x_{21}=0.25$. This means according to (8.22) that $\alpha_{11}=\alpha_{12}=0.5$, $\alpha_{21}=1$. Therefore, the optimal strategy consists of mixing in state E_1 two available CAs with equal probabilities. Note that "pure" strategies give the following values for the average costs and the stationary probabilities:

$$g(1,1)=\frac{\omega_1(1)+\omega_2(1)}{\nu_1(1)+\nu_2(1)}=-1.67, \qquad P_2=\frac{\nu_2(1)}{\nu_1(1)+\nu_2(1)}=\frac{2}{3}.$$

$$g(2,1)=\frac{\omega_1(2)+\omega_2(1)}{\nu_1(2)+\nu_2(1)}=-1, \qquad P_2=\frac{\nu_2(1)}{\nu_1(2)+\nu_2(1)}=\frac{2}{5}.$$

The mixing raised the cost rate from -1.67 up to -1.25 but provided the inequality $P_2\leqslant 0.5$.

8.6. Duality in a Problem without Constraints

This part of the section demonstrates that the problem dual to the following problem:

$$\min \bar{x}\Omega^T, \tag{8.27}$$

$$\mathbf{B}\bar{x}^T = t^T,$$

$$\bar{x} \geq 0, \tag{8.28}$$

leads to relations used in Howard's algorithm. This point seems worth examining because this algorithm was introduced without any proof and its origin might seem enigmatic to the reader.

According to the well-known duality theory in the LP (e.g. [85, Chapter 2]), $(n+1)$ variables $w_1, \ldots, w_n, z$, appear in the dual problem. Constraints in this problem have the form:

$$w_i + z\nu_i(r) \leq \omega_i(r) + \sum_{j=1}^{n} w_j p_{ij}(r); \qquad r = \overline{1, K}, \quad i = \overline{1, n} \tag{8.29}$$

and the objective is to maximize

$$0 \cdot w_1 + 0 \cdot w_2 + \cdots + 0 \cdot w_n + z.$$

Now, let us compare (8.29) and the expression (3.49) from Chapter 1 for the test quantity $z(\sigma_i)$. We see that

$$z \leq \frac{\omega_i(r) + \sum_{j=1}^{n} p_{ij}(r) w_j - w_i}{\nu_i(r)}; \qquad i = \overline{1, n}, \quad r = \overline{1, K}. \tag{8.30}$$

Theorem 1 asserts that in the problem under consideration there is an optimal solution such that every group of variables $x_{i1}, \ldots, x_{ik}$ contains exactly one nonzero, positive elements, say, x_{i1}. Now let us make use of slackness conditions:

If $x_{i1} > 0$, then the corresponding restriction in (8.30) becomes an *equality*. In other words, we shall have equalities in (8.30) for all i and for all $r = 1$ and inequalities for the other $r \neq 1$. But this is exactly the same situation we get after applying Howard's algorithm. Indeed, if the optimal strategy $\boldsymbol{\sigma}^*$ was found, then for it we have

$$z(\sigma_i^*) = \frac{\omega_i(\sigma_i^*) + \sum p_{ij}(\sigma_i^*) w_j - w_i(\sigma_i^*)}{\nu_i(\sigma_i^*)},$$

and at the same time for every other strategy $\boldsymbol{\sigma}^{**}$, according to step 3 of the algorithm,

$$\frac{\omega_i(\sigma^{**}) + \sum p_{ij}(\sigma_i^{**})w_j - w_i(\sigma_i^{**})}{\nu_i(\sigma_i^{**})} \geqslant z(\sigma_i^{**}).$$

Eventually, according to the main property of the dual LP problem, the values of objective functions are the same for the primary and dual problem, i.e.

$$\max z = z^* = \min \bar{x}\Omega^T = \min \bar{g}.$$

Of course, there exists a direct proof that Howard's algorithm provides the minimal cost rate and determines the optimal strategy (see [46] and [48]). This proof is based on the comparison of two rates g_1 and g_2, corresponding to two consecutive iterations. The reader is referred to the example at the end of Section 7 to trace the changes in the value g in each consecutive iteration cycle.

9. Controlled Markov Chain with Randomly Chosen Set of Control Actions and Control-limit Strategies

9.1. An Example and the Formal Model

Let us start with a discussion of a preventive maintenance model proposed in the book of Jorgenson, McCall and Radner [51, Chapter 3].

The equipment has Q parts, all of which are monitored. The main part numbered Q has an increasing failure rate while the remaining (auxiliary) parts have constant failure rates λ_i, $i = 1, \ldots, Q-1$. The cost of replacing an auxiliary part q is C_q. Since they fail exponentially, they are never replaced before failure. PR and ER of the main part cost C_{QQ} and C_Q^* correspondingly. When an auxiliary part q fails, there is also a possibility of carrying out its ER together with the PR of the main part, which results in cost C_{qQ}. This combined action costs less than both actions performed separately, i.e. $C_{qQ} < C_q + C_Q$. The problem is to find an optimal replacement policy which provides the minimal expected cost rate for an infinite time horizon. It will be assumed also that all repair actions are performed instantaneously.

A reasonable discrete-time version of this problem which is convenient for our further treatment would be the following. The system is inspected periodically; the interval between inspections is equal to $\Delta = 1$. There are constant probabilities p_q that the inspection will reveal a failure of an auxiliary element q. If at inspection the age of the main nonfailed part is

$\tau = i$, then the next inspection can reveal its failure with the probability

$$\rho(i) = (F(i+1) - F(i))/(1 - F(i)),$$

where $F(t)$ is the lifetime DF for this part. It is also supposed that at any inspection, only one failure can appear. That is, the probability of simultaneous appearance of more than one failure will be negligibly small.

Now we introduce the state of a system ξ_t according to the age of the main part: $\xi_t = E_i$ if on inspection the age will be equal to $\tau = i\Delta$. E_0 refers to a brand new part. In order to avoid some formal complications, it will be assumed that the age cannot exceed some maximal level n (i.e. E_n is a failure state).

Using our terminology, we can say that for the process being in the state E_i $(i \neq n)$ in the case of the failure of an auxiliary part q, there is a choice between two CAs: repair only the failed part (action 0) or combine its repair with the PM of the main part (action 1). The principal difference between this case and our control scheme proposed throughout our exposition is the following: in the "standard" control situation we have a set S_i of CAs available when the state E_i is observed. Here, there are several sets S_{iq}, $q = 1, \ldots, Q$, which are offered to the "ruler of the process" by some independent, external random mechanism. So, in the state E_i, $i < n$, the auxiliary part of the system generates randomly, with probability p_q, the set S_{iq} consisting of two actions 0 and 1; $1 - \Sigma_{q=1}^{Q} p_q$ is the probability that the set of CAs S_{iQ} will consist of actions "do nothing" or "do the PM". For the failure state E_n, we assume that there is always only one CA – "do the ER". Despite the fact that all sets S_{iq} for $q = 1, \ldots, Q-1$, contain the same CAs (0 and 1), the costs associated with them are quite different and in terms of control theory these sets must be considered as different.

The above situation suggests introducing the following formal model.

For every state E_i, at time m, a collection of sets $S_{iq}(m)$ and a collection of probabilities $\beta_{iq}(m)$, $i = 1, \ldots, n$; $q = 1, \ldots, Q$, are given. Every S_{iq} represents a set of CAs available at state E_i, at time m. $\beta_{iq}(m)$ is the probability that exactly this set will be proposed to the ruler of the process by some external source. The control goes on in the following way: state E_i is observed and set S_{iq} "is received" as an actual set of control alternatives. Later on, the best CA within this set has to be chosen in order to minimize the total costs. As usual, with every CA $\sigma_{iq}(m) \in S_{iq}(m)$, certain one-step costs $\omega_i(\sigma_{iq}(m))$ and transition probabilities $p_{ij}(\sigma_{iq}(m))$ are associated.

The principal recurrence relation for total costs in the case of finite-step process is given by the following obvious expression:

$$V_i(m+1) = \sum_{q=1}^{Q} \beta_{iq}(m+1) \left\{ \min_{\sigma_{iq}(m+1)\in S_{iq}(m+1)} \left[\omega_i(\sigma_{iq}(m+1)) + \sum_{j=1}^{n} p_{ij}(\sigma_{iq}(m+1)) V_j(m) \right] \right\}. \qquad i=1,\dots,n \tag{9.1}$$

The scheme proposed will be referred further as *random-ca-set-model* (RCS-model). The concept of RCS-model is in some sense close to the concept of *random Markovian environment* also proposed in the cited book (Chapter 3). Our nearest goal is to investigate the structure of the optimal strategy when some special assumptions concerning the stochastic behaviour of the system itself and sets S_{iq} will be made.

9.2. Optimal Strategies in RCS-model when S_{iq} Contain only Actions $E_i \to E_i$ and $E_i \to E_0$

In this paragraph we shall deal with an important special case of RCS-model when the "active" CA is realized by an immediate shift of type $E_i \to E_0$. We make the following formal assumptions:

(1) The probabilities $\beta_{iq}(m)$ do not depend on m and i: $\beta_{iq}(m) \equiv \beta_q$. The sets $S_{iq}(m)$ do not depend on the number of the step m: $S_{iq}(m) = S_{iq}$.

(2) The controlled process ξ_t is a Markov chain with states E_i, $i = 0, \dots, n$.

(3) Every set S_{iq}, $i = 0, \dots, n-1$, consists of actions $E_i \to E_i$ (denoted as 0) and $E_i \to E_0$ (denoted as 1). The set S_{nq} contains only one CA $E_n \to E_0$ (denoted as 1).

(4) Transition probabilities of the process ξ_t under different CAs are given by the relations:

$$\begin{aligned} &\boldsymbol{P}\{\xi_{t+1} = E_j \mid \xi_t = E_i\,;0\} = p_{ij}, \\ &\boldsymbol{P}\{\xi_{t+1} = E_j \mid \xi_t = E_i\,;1\} = p_{0j}. \end{aligned} \tag{9.2}$$

(5) The one-step costs associated with different CAs are equal to

$$\omega_i(\sigma_{iq}) = \begin{cases} \omega_i + C_q, & \text{if } \sigma_{iq} = 0; \quad i = 0, \ldots, n-1; \\ \omega_0 + C_{qQ}, & \text{if } \sigma_{iq} = 1; \quad q = 1, \ldots, Q-1. \end{cases}$$

$$\omega_n(\sigma_{iq}) \equiv \omega_n = C_Q^* + \omega_0; \tag{9.3}$$

$$\omega_i(\sigma_{iQ}) = \begin{cases} \omega_i, & \text{if } \sigma_{iQ} = 0; \\ \omega_0 + C_{QQ}, & \text{if } \sigma_{iQ} = 1; \end{cases} \quad i = 0, \ldots, n-1.$$

(6) The values ω_i, C_q, C_{qQ}, C_Q^* in (9.3) are positive and have the following properties:

$$\begin{aligned} &\omega_i \uparrow (i); \\ &0 < C_{qQ} - C_q \uparrow (q); \\ &C_{qQ} < C_q + C_{QQ}; \\ &C_{qQ} \leq C_Q^*. \end{aligned} \tag{9.4}$$

The symbol $f_i \uparrow (i)$ means that f_i is nondecreasing in respect to i.

(7) The transition probabilities p_{ij} of the Markov chain have an "increasing hazard-rate" property:

$$r_i = \sum_{j=l}^{n} p_{ij} \uparrow (i) \text{ for all } l. \tag{9.5}$$

(8) At every step of the process there is a nonzero stopping probability α, $0 < \alpha < 1$.

Before starting our formal reasoning, let us comment on the above conditions (1)–(8). (1) means that the "external generator" of control sets works in stationary and space-homogeneous fashion. (4) represents the influence of CA on the probabilistic behaviour of the system. The values C_{qQ} might be interpreted as costs for shifting the process when the set S_{iq} is available. The first formula in (9.4) represents the increase of one-step costs when the system approaches the "failure state" E_n. The second property in (9.4) may be always satisfied by an appropriate numeration of q. The third relation in (9.4) is introduced in order to apply the results of this model to the example described in 9.1. The condition (7) reflects natural properties of a system subjected to aging: the probability to reach the set of "dangerous" states $\{E_l, \ldots, E_n\}$ from state E_i increases with i. In the case of a single part where only transitions $E_i \to E_i$ and $E_i \to E_n$ may occur, (9.5) means that this part has an increasing failure rate. Markov chains having property (9.5) were studied by Barlow and Proschan [10] and Derman [29]. Lastly, the property (8) is of a formal nature and serves for transition from a finite-step process to an infinite-step one.

Now let us write an expression for minimal total expected costs in a one-step process. In accordance with the above suppositions, the desired formulas are:

$$\underset{i<n}{V_i(1)} = \sum_{q=1}^{Q-1} \beta_q\{\min[C_q + \omega_i;\, C_{qQ} + \omega_0]\} + \beta_Q\{\min[\omega_i;\, C_{QQ} + \omega_0]\}, \tag{9.6}$$

$$V_n(1) = \sum_{q=1}^{Q} \beta_q[C_Q^* + \omega_0] = C_Q^* + \omega_0. \tag{9.7}$$

The relation (9.6) might be rewritten in a more convenient form:

$$V_i(1) = \sum_{q=1}^{Q-1} \beta_q C_q + \sum_{q=1}^{Q-1} \beta_q\{\min[\omega_i;\, C_{qQ} - C_q + \omega_0]\}$$
$$+ \beta_Q\{\min[\omega_i;\, C_{QQ} + \omega_0]\}. \tag{9.8}$$

Lemma 1. $V_i(1) \uparrow (i)$, $i = 0, \ldots, n$.

Proof. For $i < n$ the statement is true because every expression in braces { } is nondecreasing with i according to (9.4), first formula. $V_n(1) \geq V_i(1)$ because of the last inequality in (9.4). #

Lemma 2. *For any sequence* $V_i \uparrow (i)$,

$$\sum_{j=0}^{n} p_{ij} V_j \uparrow (i). \tag{9.9}$$

Proof. Follows immediately from the representation

$$\sum_{j=0}^{n} p_{ij} V_j = \sum_{j=0}^{n} p_{ij} V_0 + \sum_{j=1}^{n} p_{ij}[V_1 - V_0] + \cdots + \sum_{j=n}^{n} p_{ij}[V_n - V_{n-1}]. \quad \#$$

Denote by $V_i(m+1;\alpha)$ the minimal total expected cost for state E_i in a $(m+1)$-step process.

Theorem 1. *For every* $m \geq 0$, $V_i(m+1;\alpha) \uparrow (i)$.

Proof. The statement is true for $m = 0$. Suppose it is true for some $m > 0$. The expressions for $V_i(m+1;\alpha)$ written in a form similar to (9.8) and (9.7) are

$$\underset{i<n}{V_i(m+1;\alpha)} = \sum_{q=1}^{Q-1} \beta_q C_q + \sum_{q=1}^{Q-1} \beta_q \Big\{ \min \Big[\omega_i + (1-\alpha) \sum_{j=0}^{n} p_{ij} V_j(m;\alpha) \Big], \Big[C_{qQ} - C_q + \omega_0 + (1-\alpha) \sum_{j=0}^{n} p_{0j} V_j(m;\alpha) \Big] \Big\}; \tag{9.10}$$

$$V_n(m+1;\alpha) = \sum_{q=1}^{Q} \beta_q \Big[C_Q^* + \omega_0 + (1-\alpha) \sum_{j=0}^{n} p_{0j} V_j(m;\alpha) \Big].$$

According to our supposition, Lemma 2 and the property of ω_i, the first term in braces { } increases with i. Therefore, $V_i(m+1;\alpha) \uparrow (i)$ for $i = 0, \ldots, n-1$. $V_n(m+1;\ \alpha) \geq V_i(m+1;\ \alpha)$ because $C_Q^* > C_{qQ} - C_q$ (see the last two inequalities in (9.4)). #

Now note that for every fixed α, $0 < \alpha < 1$, there must be a finite $\lim_{m\to\infty} V_i(m+1;\ \alpha) = V_i(\alpha)$ because $V_i(m;\alpha)$ are bounded and $V_i(m;\alpha) \uparrow (m)$. Setting $m \to \infty$ in (9.10), one obtains:

$$\underset{i<n}{V_i(\alpha)} = \sum_{q=1}^{Q-1} \beta_q C_q + \sum_{q=1}^{Q-1} \beta_q \Big\{ \min \Big[\omega_i + (1-\alpha) \sum_{j=0}^{n} p_{ij} V_j(\alpha) \Big], \Big[C_{qQ} - C_q + \omega_0 + (1-\alpha) \sum_{j=0}^{n} p_{0j} V_j(\alpha) \Big] \Big\}; \tag{9.11}$$

$$V_n(\alpha) = C_Q^* + \omega_0 + (1-\alpha) \sum_{j=0}^{n} p_{0j} V_j(\alpha).$$

From the property $V_i(m;\alpha) \uparrow (i)$, it follows now that also $V_i(\alpha) \uparrow (i)$. Denote

$$q^*(i;\alpha) = \begin{cases} \max q \text{ for which } \Big[C_{qQ} - C_q + \omega_0 + (1-\alpha) \sum_{j=0}^{n} p_{0j} V_j(\alpha) \Big] \\ \qquad < \Big[\omega_i + (1-\alpha) \sum_{j=0}^{n} p_{ij} V_j(\alpha) \Big], \\ \qquad \text{if such } q \text{ exists } (1 \leq q \leq Q); \\ 0 \quad \text{otherwise.} \end{cases} \tag{9.12}$$

Theorem 2. (i) *The optimal strategy in an infinite-step process with stopping probability* $0 < \alpha < 1$ *prescribes in state* E_i *performing action* 1 *for sets* S_q *if* $q \leq q^*(i;\alpha)$ *and action* 0 *for sets* S_q *if* $q > q^*(i;\alpha)$. (ii) $q^*(i;\alpha) \uparrow (i)$.

Proof. (i) This statement follows immediately from the definition of $q^*(i;\alpha)$. If $q^*(i;\alpha)=0$, this means that for all S_q the action 0 is optimal. (ii) follows from the definition of $q^*(i;\alpha)$, the above stated property of V_i and Lemma 2. #

The structure of the optimal strategy stated in Theorem 2 might be illustrated in terms of the example given in 9.1. Let E_i denote the actual age of the main part; the set S_q appears if the qth auxiliary part fails ($q=1,\ldots,Q-1$), action 0 corresponds to ER of the failed part only and action 1, to the combined ER of the auxiliary part and PR of the main part. The optimal strategy prescribes for age E_i of the main part carrying out the combined action if and only if an auxiliary part having number q, $q<q^*(i;\alpha)$, has failed. If we remember that auxiliary parts are numbered according to the increase of the difference $C_{qQ}-C_q$, we come to the following conclusion: it is expedient to carry out a combined action if the additional cost for uniting ER and PR is low enough. The second part of Theorem 2 asserts the following "threshold" property of the optimal strategy. Assume that for the system with age E_i it is expedient to carry out a combined repair in the case of the failure of the auxiliary part q. Then, if this part q fails when the actual age E_j of the main part is greater ($j>i$), then the combined action also is optimal. In other words, for every auxiliary part q there exists a "critical" age i_q of the main part such that for all $i\geqslant i_q$ the combined action is expedient. This property of the optimal strategy was stated in the above cited book [51] by means of a different technique.

It now remains to extend the results of Theorem 2 to the case of expected costs per unit time. This might be performed by using following standard arguments (see, for example, Derman [29, Chapter 9]).

From the finiteness of the number of stationary strategies, it follows that there exists a sequence $\alpha_k\to 0$ such that for all $k>k_0$ the optimal strategy remains the same. Denote it by $\boldsymbol{\sigma}^*$ and consider relation (9.12) for $0<\alpha<\alpha_{k_0}$. Assume that our process has only one ergodic class of nonperiodic states. It is known that $V_i(\alpha;\boldsymbol{\sigma}^*)$ have the following asymptotic behaviour (see, for example, Jewell [50]):

$$V_i(\alpha_k;\boldsymbol{\sigma}^*) \underset{k\to\infty}{=} g(\boldsymbol{\sigma}^*)/\alpha_k + w_i + o(1). \tag{9.13}$$

Substituting this relation into (9.12), we obtain after simple transformations:

$$q^*(i;\alpha_k)=\left\{\max q: C_{qQ}-C_q+\omega_0+\sum_{j=0}^{n} p_{0j}w_j<\omega_i+\sum_{j=0}^{n} p_{ij}w_j+o(\alpha_k)\right\}. \tag{9.14}$$

Denote $\lim_{\alpha_k\to 0} q^*(i;\alpha_k)=q^*(i)$. This value defines the optimal strategy for all $\alpha<\alpha_{K_0}$. It is clear that $q^*(i)\uparrow(i)$. At last, from the asymptotic relation (9.13) it follows that the strategy $\boldsymbol{\sigma}^*$ remains optimal also in respect to expected cost per unit time.

Therefore, the strategy which minimizes the expected cost per unit time prescribes carrying out action 1 in state E_i if and only if the set S_q is available, where $q\leqslant q^*(i)$; besides, the values $q^*(i)\uparrow(i)$.

9.3. *Control-limit Strategies in the Case of $Q=1$*

At this point we consider a periodically observed Markov chain controlled by "shifts" of type $E_i\to E_{\varphi(i)}$ and prove that under certain circumstances the optimal control rule is that of *control-limit* type: $\varphi(i)=i$ for $i<i^*$ and $\varphi(i)=1$ for $i\geqslant i^*$. This consideration follows Derman [29] and Kolesar [55]. Afterwards, an outline of the extension of these results to a semi-Markov process will be given.

Suppose that a Markov chain with costs is considered. The following properties are postulated:

(a) $\omega_i\uparrow(i)$ – the increase with i for one-step costs;
(b) $\sum_{j=l}^{n} p_{ij}\uparrow(i)$ for all l – increasing hazard-rate property;
(c) $1-\alpha$ is the probability of stopping at an arbitrary step;
(d) $d>0$ is the penalty for the shift of the process.

Let $V_i(m;\alpha)$ denote the total minimal expected cost accumulated during m steps when process starts in E_i. These values satisfy the recurrence relation:

$$V_i(m+1;\alpha)=\min_{1\leqslant k\leqslant n}\left[d(1-\delta_{ik})+\omega_k+(1-\alpha)\sum_{j=1}^{n} p_{kj}V_j(m;\alpha)\right]. \tag{9.15}$$

Here δ_{ik} is the Kroneker symbol and k represents the CA of type $E_i\to E_k$.

It is easy to check that $V_i(1;\alpha)\uparrow(i)$. Suppose that for an arbitrary m, $V_i(m;\alpha)\uparrow(i)$. It is easy to conclude from (9.15) by help of (a) and (b) that in every state only two CAs $E_i\to E_1$ and $E_i\to E_i$ are worth considering. In other words,

$$V_i(m+1;\alpha) = \min\left[d+\omega_1+(1-\alpha)\sum_{j=1}^{n} p_{ij}V_j(m;\alpha);\right.$$

$$\left.\omega_i+(1-\alpha)\sum_{j=1}^{n} p_{ij}V_j\right]. \qquad (9.16)$$

From this relation, in turn, it follows also that $V_i(m+1;\alpha)\uparrow(i)$. Proceeding in a similar way to the previous point, we consider a limiting case when $m\to\infty$ and find out that $\lim_{m\to\infty} V_i(m;\alpha) = V_i(\alpha)$ and $V_i(\alpha)\uparrow(i)$. $V_i(\alpha)$ satisfy the relation:

$$V_i(\alpha) = \min_{k=1,i}\left[d(1-\delta_{ik})+\omega_i+(1-\alpha)\sum_{j=1}^{n} p_{kj}V_j(\alpha)\right]. \qquad (9.17)$$

Assume that there exists a number $k(\alpha)$, $1<k(\alpha)\leqslant n$, such that

$$k(\alpha) = \left\{\min k : \omega_k+(1-\alpha)\sum_{j=1}^{n} p_{kj}V_j(\alpha) \geqslant d+\omega_1\right.$$

$$\left.+(1-\alpha)\sum_{j=1}^{n} p_{1j}V_j(\alpha)\right\}. \qquad (9.18)$$

Then the optimal strategy is of *control-limit* type: in states E_i with $i<k(\alpha)$ "do nothing" ($E_i\to E_i$) and in states E_i with $i\geqslant k(\alpha)$, shift the process into state E_1. If such a threshold $k(\alpha)$ cannot be found, that is for all k,

$$\omega_k+\sum_{j=1}^{n} p_{kj}V_j(\alpha) < d+\omega_1+\sum_{j=1}^{n} p_{1j}V_j(\alpha),$$

then for all states the optimal CA is "do nothing".

The above property of the optimal strategy is in an agreement with common sense because our suppositions (a) and (b) mean that states E_i with larger number i became both "more costly" and "more dangerous".

By applying standard arguments, one can conclude that the optimal strategy is of *control-limit* type also in respect to the expected cost per unit time.

We proceed with an outline of how similar results can be obtained for the case of a semi-Markov process. The details of proofs are given in [40].

We consider a controlled SMP with CAs realized by shifts of type $E_i\to E_{\varphi(i)}$. A continuous discount factor $e^{-\alpha\tau}$ is introduced (see Section 3.6, Chapter 1). The corresponding recurrence relations for the minimal total costs are the following:

$$v_i(\alpha) = \min_{1 \le k \le n} \left[d(1-\delta_{ik}) + \omega_k(\alpha) + \sum_{j=1}^{n} p_{kj} \int_0^\infty e^{-\alpha\tau} v_j(\alpha)\, dF_{kj}(\tau) \right]. \quad (9.19)$$

If we expand $e^{-\alpha\tau} = 1 - \alpha\tau + \alpha^2 t^2 O(1)$ for $\alpha \to 0$ and substitute it in above relation, we obtain the formula:

$$v_i(\alpha) = \min_{1 \le k \le n} \left[d(1-\delta_{ik}) + \omega_k + \sum_{j=1}^{n} p_{kj}(1 - \alpha\nu_{kj}) v_j(\alpha) + \Psi(\alpha) \right], \quad (9.20)$$

where $\Psi(\alpha)_{\alpha\to 0} = O(\alpha)$ and $\nu_{kj} = \int_0^\infty t\, dF_{kj}$ (the expected transition times from E_k to E_j).

(9.20) is similar to (9.17) and it can be considered as a recurrence relation for a special Markov chain which has the transition probabilities

$$\tilde{p}_{kj}(\alpha) = p_{kj}(1 - \alpha\nu_{kj}). \quad (9.21)$$

Assume that this chain has the following property: there exists an $\alpha^* > 0$ such that for all α, $0 < \alpha \le \alpha^*$, the sum

$$\sum_{j=m}^{n} \tilde{p}_{ij}(\alpha)$$

is increasing with i for all m.

This property is a natural extension of an increasing hazard-rate property to a SMP. It can be proved that it guarantees the optimality of the *control-limit* strategy in the SMP in respect to expected cost per unit time.

10. Problems and Exercises

10.1. Optimal Preventive Maintenance in a System with a Standby which is Operating for a Finite Time – Optimization in Respect to the Maximum of Nonidle Time

In the problem considered in Section 2, we introduced the following times for the system's service: m_s – time for an inspection; m_a – time for an ER; m'_p – time for the repair of type $E_2^* \to E_0$; m''_p – for the repair $E_1 \to E_0$ or $E_2^* \to E_1$. It is necessary to find a service policy which maximizes the average time the system spent in states E_0 and E_1. All other conditions of the problem remain the same.

Derivation of recurrence relations

Let us define the following one-step returns for states E_0, E_1 and E_2^*:

$$\omega_0 = P_{00}(K_0)K_0 + P_{01}(K_0)K_0 + \sum_{z=1}^{K_0} P_{02}(z)z.$$

(Here K_0 is the time interval from the end of the repair until the beginning of the next inspection.)

$$\omega_1 = \begin{cases} \omega_0, & \text{if the repair } E_1 \to E_0 \text{ was made;} \\ \omega_1' = P_{11}(K_1)K_1 + \Sigma_1^{K_1} P_{12}(z)z, & \text{if in state } E_1 \text{ nothing has been done and the next inspection is planned after time } K_1. \end{cases}$$

$$\omega_2 = \begin{cases} \omega_0, & \text{if in state } E_2^* \text{ the repair } E_2^* \to E_0 \text{ was made;} \\ \omega_1, & \text{if the repair } E_2^* \to E_1 \text{ was made.} \end{cases}$$

Let the moment $t = N - (m+1)$ be the end of the inspection, as a result of which state E_i is discovered. Denote by $v_i(m+1)$ the total average working time during the interval $[N-(m+1), N]$ if the optimal service strategy is used.

The values $v_i(k)$ satisfy the following recurrence relations:

$$\begin{aligned} v_0(m+1) = \omega_0 &+ P_{00}(K_0)v_0(m+1-K_0-m_s) \\ &+ P_{01}(K_0)v_1(m+1-K_0-m_s) \\ &+ P_{02}(K_0)v_2(m+1-K_0-m_s); \end{aligned}$$

$$\begin{aligned} v_1(m+1) = \max\{&\omega_0 + P_{00}(K_0)v_0(m+1-K_0-m_p''-m_s) \\ &+ P_{01}(K_0)v_1(m+1-K_0-m_p''-m_s); \\ &\omega_1' + P_{11}(K_1)v_1(m+1-K_1-m_s) + P_{12}(K_1)v_2(m+1-K_1-m_s)\}; \end{aligned}$$

$$\begin{aligned} v_2(m+1) = \max\{&\omega_0 + P_{00}(K_0)v_0(m+1-K_0-m_a-m_p'-m_s) \\ &+ P_{01}(K_0)v_1(m+1-K_0-m_a-m_p'-m_s) \\ &+ P_{02}(K_0)v_2(m+1-K_0-m_a-m_p'-m_s); \\ &\omega_1' + P_{11}(K_1)v_1(m+1-K_1-m_a-m_p''-m_s) \\ &\qquad + P_{12}(K_1)v_2(m+1-K_1-m_a-m_p''-m_s)\}. \end{aligned}$$

These relations are valid if the arguments of functions v_i are nonnegative. Otherwise, obvious changes must be made in the above expression (see Section 2).

10.2. Preventive Maintenance Using a Prognostic Parameter with Randomly Chosen Critical Level

Let the maximum in $g(h)$ (see (6.6)) be attained at the level h^*:

$$\max_h g(h) = g(h^*).$$

Assume that the critical level $\mathbf{h}$ for the prognostic parameter is chosen before the beginning of the cycle of work according to a DF $H(x) = P\{\mathbf{h} \leq x\}$, and that the corresponding operational readiness is equal to $g(\mathbf{h})$. Prove that $g(h^*) \geq g(\mathbf{h})$.

Solution. When a random level is selected, the OR is equal to the ratio

$$\int_0^\infty \Psi(h)\, \mathrm{d}H(h) \Big/ \int_0^\infty \varphi(h)\, \mathrm{d}H(h),$$

where $\Psi(h)$ and $\varphi(h)$ denote numerator and denominator in (6.6). From the definition of h^* it follows that

$$\frac{\Psi(h^*)}{\varphi(h^*)} \geq \frac{\Psi(h)}{\varphi(h)},$$

and, therefore, $\Psi(h^*)\varphi(h) \geq \varphi(h^*)\Psi(h)$. Multiplying both sides of that inequality by $\mathrm{d}H(x)$ and integrating from 0 up to ∞, we obtain the desired inequality. The idea of this proof follows [8].

10.3. Service Strategy when there is no Signalling of Failure

The problem discussed in Section 7 is modified as follows. All $K_i = K = \text{Const}$. No information is available about the failure. The system's state is revealed only at the inspection.

Write the expression for the average cost per unit time for a given strategy.

10.4. Additional Penalties in Problems Discussed in Sections 3 and 7

Write the expressions for one-step costs if there are additional penalties proportional to holding times in states E_i for the process $\eta(t)$.

Solution. To calculate the one-step costs, it is necessary to know the average holding times $l_{j,\varphi(i)}$, $j = 0, \ldots, n-1$, for the process $\eta(t)$ if the CA $\sigma_i = \{\varphi(i), T_{\varphi(i)}\}$ is selected.

Define the random variable

$$I_j(t) = \begin{cases} 1, & \text{if } \eta(t) = E_j, \\ 0, & \text{if } \eta(t) \neq E_j. \end{cases}$$

Then

$$l_{j,\varphi(i)} = \boldsymbol{M}\left\{ \int_0^{T_{\varphi(i)}} I_j(t)\,\mathrm{d}t \right\} = \int_0^{T_{\varphi(i)}} \boldsymbol{M}\{I_j(t)\}\,\mathrm{d}t$$

$$= \int_0^{T_{\varphi(i)}} \boldsymbol{P}\{\eta(t) = E_j \mid \eta(0) = E_{\varphi(i)}\}\,\mathrm{d}t = \int_0^{T_{\varphi(i)}} P_{\varphi(i),j}(t)\,\mathrm{d}t.$$

The total expected one-step cost for state E_i is equal to

$$\omega_i(\sigma_i) = c_{i,\varphi(i)} + \sum_{j=0}^{n-1} l_{j,\varphi(i)}\, d_j.$$

10.5. Continuation of 10.4.

In the problem considered in Section 7, it is assumed that the information about the failure comes with the probability γ, $0 < \gamma < 1$. If the failure occurred and no information had been received, it will be revealed when the next inspection is made. For every time unit the process $\eta(t)$ spends in state E_1, an additional penalty d_i, $i = \overline{0, n}$ is counted.

Derive the expressions for transition probabilities $p_{ij}(\sigma_i)$ and one-step costs $\omega_i(\sigma_i)$.

10.6. Strategies of type $\sigma_i = (\varphi(i), T_{i,\varphi(i)})$ in the problem considered in Section 7

It is necessary to show that the optimal strategy has the following property: $T_{i,\varphi(i)} = T_{\varphi(i)}$, i.e. the period of the inspection does not depend on the state initially observed in the process ξ_t and actually depends only on the state in which the system has started to work.

Solution. Let $\boldsymbol{\sigma}^* = (\sigma_1^*, \ldots, \sigma_n^*)$ be the optimal strategy. Let us write the system of equations (7.17) for that strategy and solve it in respect to $w_0, \ldots, w_n, g$. Then we obtain

$$g^* = (\omega_i(\sigma_i^*) + \sum_j p_{ij}(\sigma_i^*) w_j^* - w_i^*)/\nu_i(\sigma_i^*).$$

The comparison with the test ratios in Howard's algorithm shows that for some other CA, $\sigma_i^{**} \neq \sigma_i^*$,

$$g^* < \left(\omega_i(\sigma_i^{**}) + \sum_j p_{ij}(\sigma_i^{**}) w_j^* - w_i^*\right) \Big/ \nu_i(\sigma_i^{**}).$$

Comparing these two expressions, we obtain

$$\omega_i(\sigma_i^*) + \sum_j p_{ij}(\sigma_i^*) w_j^* - g^* \nu_i(\sigma_i^*)$$
$$< \omega_i(\sigma_i^{**}) + \sum_j p_{ij}(\sigma_i^{**}) w_j^* - g^* \nu_i(\sigma_i^{**}).$$

Let us take $\sigma_i^* = (\varphi(i), T_1)$, $\sigma_i^{**} = (\varphi(i), T_2)$, $T_2 \neq T_1$. Then the last inequality can be rewritten in the form

$$C_{i,\varphi(i)} + \omega_{\varphi(i)}(T_1) + \sum_j p_{\varphi(i),j}(T_1) w_j^* - g^* \nu_{\varphi(i)}(T_1)$$
$$\leqslant C_{i,\varphi(i)} + \omega_{\varphi(i)}(T_2) + \sum_j p_{\varphi(i),j}(T_2) w_j^* - g^* \nu_{\varphi(i)}(T_2). \qquad (*)$$

Now, let us assume that for the state E_k, $k \neq i$, the optimal CA will be $\sigma_k^* = (\varphi(i), T_2)$. Then

$$C_{k,\varphi(i)} + \omega_{\varphi(i)}(T_2) + \sum_j p_{\varphi(i),j}(T_2) w_j^* - g^* \nu_{\varphi(i)}(T_2)$$
$$< C_{k,\varphi(i)} + \omega_{\varphi(i)}(T_1) + \sum_j p_{\varphi(i),j}(T_1) w_j^* - g^* \nu_{\varphi(i)}(T_1). \qquad (**)$$

But (*) contradicts (**). That excludes the possibility $T_2 \neq T_1$. Therefore, $\sigma_k^* = (\varphi(i), T_1)$.

Chapter 4

SPECIAL QUESTIONS RELEVANT TO OPTIMAL PREVENTIVE MAINTENANCE

1. Introduction

This chapter is concerned with some questions which are important in the practical application of preventive maintenance to real systems. Every application of a formal model to practice usually entails great difficulties. This often arises when an analysis of the real situation reveals that the formal premises on which the mathematical model was based do not correspond to actual conditions. Such a situation is familiar to everyone who has dealt with practical operations research and has tried to apply theoretical models in real life circumstances.

The models of preventive maintenance considered earlier in this book have many vulnerable points. In the first place, almost all models were devoted to the maintenance of systems having one part (one device or element) subject to failure. How does one organize the optimal, or at least reasonable, service process for a system consisting of many parts or devices? Even if the statistical dependence of these parts can be ignored, we cannot ignore the "economic" dependence shown by the fact that combining preventive actions for different elements (parts, devices) usually results in lower costs or other expenses than executing the same actions separately for each element or device.

We have already touched on several models of preventive maintenance in which this "economic" dependence was more or less reflected (see, for example, the model of group preventive maintenance, Chapter 2, Section 4). Here, in Section 2, we shall simplify the replacement rule by reducing it to a periodic block replacement and at the same time, we shall go further in considering the relations between the cost of preventive maintenance and its "depth".

The attentive reader has no doubt noticed that Chapter 3 considers only systems having a *one*-dimensional parameter. This is, of course, a serious simplification of the real picture. Every engineer knows that usually he has to predict the state of certain technical systems, taking into account information about a multidimensional parameter. Thus, a decision to overhaul an engine must be based on taking into consideration several parameters: accrued operation time, fuel consumption, efficiency of the engine, etc. Formally speaking, it would be desirable to have models of preventive maintenance based on the supervision of some multidimensional parameters. However, this is a very difficult problem, especially in relation to its statistical essence. It is a very difficult task to provide satisfactory statistical description of a multidimensional random process using as a basis some rather poor data which are usually at our disposal. One way out of this difficulty is to go from a multidimensional to a one-dimensional situation, as shown in proposals represented in Section 3 of this chapter.

Thus, this chapter is designed to fill a gap in our collection of preventive maintenance models and to offer the reader a more useful and realistic approach to their utilization.

2. *Optimal Grouping of Maintenance Times for a Hierarchically Organized Multicomponent System*

2.1. General Description

Let us consider preventive maintenance for a system consisting of several statistically independent parts. Our objective is to find a rule (schedule) for performing PR for every element of this system. Solving this problem involves some specific difficulties caused by the following important circumstances.

The time or money spent for simultaneous repairs of a group of elements is essentially less than that spent on repairing each individual part (element) separately. This suggests grouping the elements while carrying out preventive repairs. Assume, for example, that we have a system consisting of two elements; an optimal preventive maintenance period is selected for each element which is considered separately. Denote these periods T_1 and T_2, $T_1 \neq T_2$. Some amount of time or money is spent, C_1 and C_2, for the repair of these elements. If the PR time moments are combined for both elements, probably the cost will be equal to a certain value x, which yields the following inequality:

$$\max(C_1, C_2) < x < C_1 + C_2.$$

Most probably, it would be more profitable to consider the system of two elements as a single unit and introduce a common period T, $T \neq T_1$, $T \neq T_2$, of periodic maintenance when both elements could be repaired together.

All the rules prescribing the maintenance of complex equipment foresee grouping the elements or parts of subsystems subject to preventive actions; such amalgamation of service actions becomes much easier when the preventive maintenance is planned periodically. This is exactly the situation we had in mind in planning block replacement. In the sequel, we shall formulate a mathematical model of a multicomponent system in which we develop the idea of optimal grouping of the elements according to their preventive maintenance periods.

2.2. *Structure of the System and Penalties*

Assume that the system consists of elements which will be denoted by $i = (i_1, \ldots, i_n)$. They constitute the set $\mathbf{I}_0$. All elements having fixed first k coordinates, $i_1 = i_1^*, \ldots, i_k = i_k^*$, $k = 1, \ldots, n-1$, constitute one subsystem of the kth rank, and will be denoted by a k-tuple $\{i_1^*, \ldots, i_k^*\}$. One separate element $i = \{i_1, \ldots, i_n\}$ is considered as a subsystem of the rank n. This dimension n of the vector i is called the order of the system.

In Figure 37, a system of the third order is plotted. Elements (1, 0, 0), (1, 1, 0), (1, 1, 1), (1, 1, 2), (1, 2, 0) and (1, 2, 1) form one subsystem {1} of the rank 1. Elements (1, 2, 0) and (1, 2, 1) form the subsystem {1, 2} of rank 2.

Every element is subjected to a periodic preventive maintenance, that is, a period τ_i for its PR is fixed. It is assumed that PR renews the element entirely. During the time when one element is being repaired, other elements are not working and their "lifetime resources" are not used. Let $\varphi_i(\tau_i)$ denote the average cost associated with the element $\mathbf{i}$ including costs for ER and PR counted over the interval $(0, \tau_i)$. For example, in block replacement, $\varphi_i(\tau_i) = M(\tau_i)c_a + c_p$, where $M(\tau_i)$ is the ME of the number of failures occurring in the interval $(0, \tau_i)$, c_a and c_p are ER and PR costs.

A special feature of the system under consideration is the presence of additional penalties corresponding to preventive repairs for subsystems of different ranks.

Let the constants $f, f_{i_1}, \ldots, f_{(i_1,\ldots,i_{n-1})}$ be given. The constant $f_{(i_1,\ldots,i_r)}$ will be considered as the rth rank penalty. It will be assumed that if at the same time a certain group $\mathbf{I} = \{i\}$ of elements is subjected to preventive

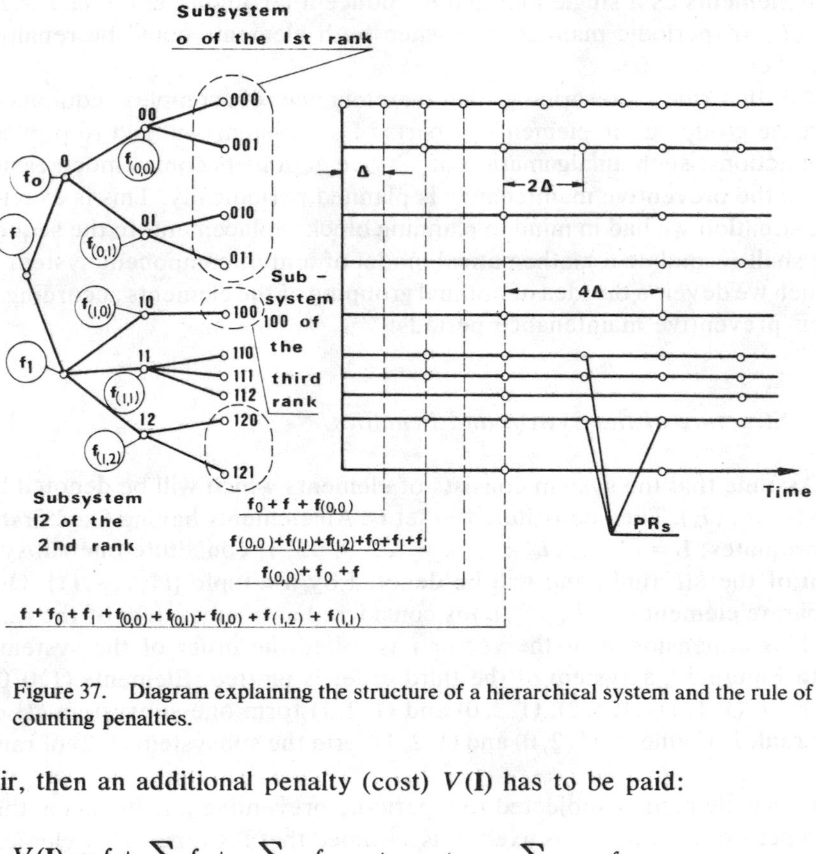

Figure 37. Diagram explaining the structure of a hierarchical system and the rule of counting penalties.

repair, then an additional penalty (cost) $V(\mathbf{I})$ has to be paid:

$$V(\mathbf{I}) = f + \sum_{j_1 \in \mathbf{I}_1} f_{j_1} + \sum_{(j_1, j_2) \in \mathbf{I}_2} f_{(j_1, j_2)} + \cdots + \sum_{(j_1, \ldots, j_{n-1}) \in \mathbf{I}_{n-1}} f_{(j_1, j_2, \ldots, j_{n-1})}. \tag{2.1}$$

In this formula, J_r indicates the set of all r-dimensional subvectors whose components coincide with corresponding first r components of vectors $i \in \mathbf{I}$. (If $i = (i_1, i_2, \ldots, i_n)$, then its subvectors are (i_1), (i_1, i_2), (i_1, i_2, i_3), etc.) We give an example explaining formula (2.1).

Let us consider a system plotted on Figure 37. We define f – the "all-system" penalty; f_0, f_1 – first-rank or "subsystems" penalties; $f_{(0,0)}$, $f_{(0,1)}$, $f_{(1,0)}$, $f_{(1,1)}$, $f_{(1,2)}$ – second-rank or "pack" penalties. Let the elements (0, 0, 0), (0, 0, 1), (1, 1, 0), (1, 1, 1), and (1, 2, 0) which constitute set $\mathbf{I}$, undergo repair simultaneously. Then $\mathbf{I}_1 = \{0, 1\}$, $\mathbf{I}_2 = \{(0, 0), (1, 1), (1, 2)\}$, and according to (2.1) the following cost must be paid:

$$V(\mathbf{I}) = f + f_0 + f_1 + f_{(0,0)} + f_{(1,1)} + f_{(1,2)}. \tag{2.1*}$$

It is not difficult to present a situation which explains the origin of all these costs. Thus, for example, the whole system might be a truck, {0} – its engine, {1} – its chassis, {0, 0} – piston device, {0, 1} – fuel supply system, {1, 0} – braking system, {1, 1} – steering device, {1, 2} – gear-box.

Then the cost f is paid every time when the truck has stopped functioning; f_0 represents the additional cost paid for testing the engine; the additional cost f_1 is paid for a general inspection of the chassis. $f_{(0,0)}$ corresponds to the cost arising from dismantling, assembling and testing the subsystem {0, 0}, etc.

The above set **I** indicates that parts (0, 0, 0) and (0, 0, 1) of the piston device, two parts (1, 1, 0) and (1, 1, 1) of the chassis and one part (1, 2, 0) of the gear-box are repaired together, at the same time. In other words, subsystems {0} and {1} and "packs" {0, 0}, {1, 1}, {1, 2} are repaired simultaneously. This explains the above expression (2.1*).

It should be pointed out that those costs which are directly connected with repaired elements are not included in the above expression. Suppose, for example, that the element (0, 0, 0) is a crank-shaft (an element of the engine). Then the specific expenditures for its repair (testing this element, changing it for a new one, repairing its bearings, etc.) are not included in (2.1) but will be incorporated in the value $\varphi_{(0,0,0)}(\tau_{(0,0,0)})$.

Let $a_1, a_2, \ldots, a_{k-1}$ be positive integers; Δ is a given time interval. It will be assumed that the intervals between PRs for every element can only be the following values:

$$t_1 = \Delta,\ t_2 = t_1 a_1, \ldots, t_k = t_{k-1} a_{k-1}. \tag{2.2}$$

For example, assume $\Delta = 10$ hours, $a_1 = a_2 = a_3 = a_4 = a_5 = a_6 = 2$. This means that the period of preventive maintenance can be only 10, 20, 40, 80, 160, 320 and 640 hours. The service regulation for many technical systems such as aircraft or trucks is based mainly on this principle. The hours correspond to the "time in the air" or to the mileage. The condition (2.2) is a contributory factor which offers the possibility of uniting at the same time the repair of many elements and therefore, helps to reduce repair costs.

Let the period of x_i of PR be appointed for the element i, $x_i \in \{t_s\}$. We want to obtain an expression for the total additional cost rate generated by all subsystems of all ranks k, $k = 1, \ldots, n-1$.

Consider a certain subsystem $\{i_1^0, \ldots, i_s^0\}$ of sth rank. The cost $f_{(i_1^0,\ldots,i_s^0)}$ must be paid every time an element of type $i = (i_1^0, \ldots, i_s^0; i_{s+1}, \ldots, i_n)$

belonging to this subsystem is repaired. Let us consider the set of all elements of this subsystem and let $\min_{\{i_s+1,\ldots,i_n\}} x_{(i_1^0,\ldots,i_s^0;\, i_s+1,\ldots,i_n)}$ be the minimal PR period within this subsystem. Then, according to the above property (2.2), the cost $f_{(i_1^0,\ldots,i_s^0)}$ will be paid only when the element with *minimal* PR periodicity within this subsystem is repaired. Therefore, the contribution of this subsystem to the total cost rate is equal to the quotient:

$$f_{(i_1^0,\ldots,i_s^0)} \Big/ \min_{\{i_s+1,\ldots,i_n\}} x_{(i_1^0,\ldots,i_s^0;\, i_s+1,\ldots,i_n)}.$$

In addition, the all-system cost f will be paid every time an element is repaired, which results in the cost rate $f/\min_i x_i$. Therefore, the total additional cost rate is equal to the sum:

$$f_\Sigma(\{x_i\}) = \frac{f}{\min\limits_i x_i} + \sum_{r\in \mathbf{I}_1} \frac{f_r}{\min\limits_{\{i_2,\ldots,i_n\}} x_{(r,i_2,\ldots,i_n)}} + \cdots + \sum_{\{i_1,\ldots,i_{n-1}\}\in \mathbf{I}_{n-1}} \frac{f_{(i_1,\ldots,i_{n-1})}}{\min\limits_{\{i_n\}} x_{(i_1,\ldots,i_{n-1},i_n)}}. \tag{2.3}$$

Next, the general expression including all costs will be obtained. If the element i has the period of PR x_i, then the total average cost per unit time for the whole system is:

$$W(\{x_i\}) = f_\Sigma(\{x_i\}) + \sum_{i\in \mathbf{I}_0} \varphi_i(x_i)/x_i. \tag{2.4}$$

The problem we are concerned with is to find values x_i which minimize $W(\{x_i\})$.

In the next part of this section we shall give an algorithm for finding the optimal collection of x_i for the system of the first order. Later, the algorithm will be extended for the system of an arbitrary order.

2.3. *Optimal Periods for the System of the First Order*

The system consists of n elements repaired with periods $x_1, \ldots, x_n$. The total average cost rate for this case is given by the formula

$$W(x_1,\ldots,x_n) = \frac{f}{\min\limits_{\{j\}} x_j} + \sum_{j=1}^{n} \varphi_j(x_j)/x_j. \tag{2.5}$$

In the sequel, it will be assumed that functions $\varphi_i(x)$ for $x>0$ are continuous, positive and satisfy the following convexity condition:

$$\alpha\varphi_i(x)+(1-\alpha)\varphi_i(y)>\varphi_i(\alpha x+(1-\alpha)y), \qquad 0<\alpha<1.$$

It is easy to show that if x_0 is the minimal value of the ratio $\varphi_i(x)/x$ on the interval $[x', x'']$, $0<x'<x_0<x''$, then $\varphi_i(x)/x$ monotonically increases to the left and to the right from the minimum point. If the point x_0 coincides with x', then this ratio increases monotonically with the increase of x; if $x_0=x''$, then $\varphi_i(x)/x$ decreases with x. To establish this, one must consider the intersection points of a line $y_1=kx$ with the convex curve $y_2=\varphi(x)$.

Assume that the optimal period t_i^* for PR is found for every element in the first order system (FOS). Let

$$\min_{t\in\{\Delta,\, a_1\Delta,\, a_2\Delta,\dots\}} \frac{\varphi_i(t)}{t}=\frac{\varphi_i(t_i^*)}{t_i^*}.$$

Let us then enumerate all elements according to the increase of values t_i^*.

To simplify the notations, we shall not write the values t_i^* themselves, but their ordinal numbers in the ordered collection $t_1<t_2<\cdots<t_K$. Therefore, the optimal periods for FOS form a n-tuple

$$u^{(0)}=(k_1,k_2,\dots,k_n), \qquad k_i\uparrow(i). \tag{2.6}$$

If, for example, $t_K=2^K\Delta$, $t_1^*=2^3\Delta$, $t_3=\cdots=t_n=2^4\Delta$, then we shall write

$$u^{(0)}=(3,4,4,\dots,4).$$

The notation $u\geqslant u^{(0)}$, $u=(u_1,\dots,u_n)$, will designate the fact that $u_i\geqslant u_i^{(0)}$. (u_i and $u_i^{(0)}$ are components of vectors u and $u^{(0)}$.)

Let D be a constant (integer). Let $u=\max(v,D)$ denote a vector which has components equal to $u_i=\max(v_i,D)$.

Note that if $u\geqslant v\geqslant u^{(0)}$, $\min_{1\leqslant i\leqslant n}u_i=\min_{1\leqslant i\leqslant n}v_i$, then $W(u)>W(v)$. This follows from the comparison of expressions (2.5) for $W(u)$ and $W(v)$: the first terms are the same, while the second term is larger for the tuple u because of the above property of the ratio $\varphi_i(x)/x$.

Let S be a set of all vectors not smaller than $u^{(0)}$:

$$S=\{u: u\geqslant u^{(0)}\}$$

and let $\tilde{u}\notin S$. Then there must be a component of $\tilde{u}$, u_r, such that $u_r^{(0)}>\tilde{u}_r$. Let

$$\tilde{u}^*=\max(\tilde{u},u^{(0)}).$$

Then $W(\tilde{u}) > W(\tilde{u}^*)$. Indeed, the increase of the rth component from $\tilde{u}_r$ up to $u_r^{(0)}$ may only lower the contribution of the rth element in formula (2.4). This is due to the property of $\varphi_r(x)/x$ and the fact that $u_r^{(0)}$ is the optimal PR period for a separately considered element r. Besides, the first term in (2.5) cannot be larger for $\tilde{u}^*$ than for $\tilde{u}$. Therefore, the optimal tuple of periodicities can be sought within the set S.

Now consider the partition of S

$$S = \bigcup_{j \geqslant k_1} S_j, \tag{2.7}$$

where

$$S_j = \{u : u \geqslant u^{(0)}, \min_i u_i = j\}.$$

In other words, S_j consists of all vectors which have a minimal component equal to j. We recall that j designates the PR period t_j and k_1 is the minimal member in the tuple $u^{(0)}$. In order to unify notations, we shall write $u^{(0)} = v^{(k_1)}$.

Now the following property will be stated: within S_j the best is the tuple

$$v^{(j)} = \max(v^{(K_1)}, j).$$

To prove this, we note that all members of S_j and $v^{(j)}$ have the same minimal component j. This provides the equality of the first terms in (2.5) for $W(v^{(j)})$ and $W(v)$, $v \in S_j$. From the definition of S_j it follows that $v \geqslant v^{(j)}$. By the stated property of the quotient $\varphi(x)/x$, the second term in (2.5) can only increase when the tuple $v^{(j)}$ is replaced by $v \in S_j$.

Therefore, the optimal tuple of periodicities in the class S can be sought within the sequence $v^{(j)}$, $j \geqslant k_1$. Loosely speaking, it is obtained from $v^{(k_1)} = u^{(0)}$ by means of a gradual increase of the minimal component. For example, let $t_1 = 2$, $t_2 = 4$, $t_3 = 8$, $t_4 = 16$, $t_5 = 32$ and $t_1^* = 4$, $t_2^* = 4$, $t_3^* = 8$, $t_4^* = 8$, $t_5^* = 32$. In our notations,

$$\begin{aligned} u^{(0)} = v^{(1)} &= (2, 2, 3, 3, 5), \\ v^{(2)} &= (3, 3, 3, 3, 5), \\ v^{(3)} &= (4, 4, 4, 4, 5), \\ v^{(4)} &= (5, 5, 5, 5, 5). \end{aligned}$$

The sequence $v^{(k)}$ will be termed as a basic one. The number of elements in it is limited by the longest PR period t_K $(k_1 \leqslant k \leqslant K)$.

Finding the optimal tuple will be made by computation and comparison of values $W(v^{(k)})$, $k \geq k_1$. The following property can simplify the search for an optimal tuple: let r be the smallest integer such that

$$W(v^{(k_1+r-1)}) \geq W(v^{(k_1+r)}) < W(v^{(k_1+r+1)}). \tag{2.8}$$

Then

$$W(v^{(k_1+r+m)}) \uparrow (m), \qquad m \geq 0. \tag{2.9}$$

From (2.8) and (2.9) it follows that the tuple $v^{(k_1+r)}$ might be considered as an optimal one.

Comparing (2.5) for $v^{(k_1+r)}$ and $v^{(k_1+r+1)}$, we obtain after simple transformations:

$$(t_{k_1+r+1} - t_{k_1+r})^{-1} \sum_{i \in R} [t_{k_1+r+1}\varphi_i(t_{k_1+r}) - t_{k_1+r}\varphi_i(t_{k_1+r+1})] + f < 0. \tag{2.10}$$

In this expression, R denotes the set of those elements for which the increase in the period of maintenance took place when the vector $v^{(k_1+r)}$ was replaced by the vector $v^{(k_1+r+1)}$. According to the definition of $v^{(k)}$, the change in periods means the increase from t_{k_1+r} to t_{k_1+r+1}, for all elements of the set R.

In order to prove that

$$W(v^{(k_1+r+1)}) < W(v^{(k_1+r+2)}), \tag{2.11}$$

it is enough to prove the following inequality:

$$(t_{k_1+r+2} - t_{k_1+r+1})^{-1} \sum_{i \in R'} [t_{k_1+r+2}\varphi_i(t_{k_1+r+1}) - t_{k_1+r+1}\varphi_i(t_{k_1+r+2})] + f < 0, \tag{2.12}$$

where R' is the set of elements for which the change of maintenance periods took place when the vector $v^{(k_1+r+1)}$ was replaced by $v^{(k_1+r+2)}$. From the construction of the basic sequence it follows that $R \subset R'$. For an element $j \in R' - R$, the corresponding components in the formula for $W(v^{(k_1+r+2)})$ are larger than the same components in the formula for $W(v^{(k_1+r+1)})$, according to the property of the ratios $\varphi(x)/x$. Therefore, it is sufficient to check (2.12) for $i \in R$ only. The inequality (2.12) follows from (2.10) if it is true that

$$\begin{aligned}(t_{k_1+r+1} - t_{k_1+r})^{-1}[t_{k_1+r+1}\varphi_i(t_{k_1+r}) - t_{k_1+r}\varphi_i(t_{k_1+r+1})] \\ > (t_{k_1+r+2} - t_{k_1+r+1})^{-1}[t_{k_1+r+2}\varphi_i(t_{k_1+r+1}) - t_{k_1+r+1}\varphi_i(t_{k_1+r+2})].\end{aligned} \tag{2.13}$$

However, this is valid because it is equivalent to the inequality

$$[t_{k_1+r+1}(t_{k_1+r+2}-t_{k_1+r+1})\varphi_i(t_{k_1+r})+t_{k_1+r+1}(t_{k_1+r+1}-t_{k_1+r})\varphi_i(t_{k_1+r+2})]$$
$$\times[t_{k_1+r}(t_{k_1+r+2}-t_{k_1+r+1})+(t_{k_1+r+1}-t_{k_1+r})t_{k_1+r+2}]^{-1}>\varphi_i(t_{k_1+r+1}),$$

following from the convexity of $\varphi(t)$.

The property (2.9) follows from this reasoning by induction.

Now we are in a position to present an algorithm for finding the optimal PR periods for the FOS.

(1) Find optimal periods t_j for all elements.
(2) Form the initial tuple $v^{(k_1)}$ and the basic sequence $\{v^{(k)}\}$.
(3) Calculate the total costs $W(v^{(k)})$ for the vectors of the basic sequence.

The smallest integer r, $r<K$, which satisfies the inequality $V(v^{(r)})<V(v^{(r+1)})$ corresponds to the optimal tuple of maintenance periods. Otherwise the tuple $v^{(K)}$ is optimal.

2.4. Optimal Periods for the Second-order System – Algorithm for the General Case

We shall omit the formal arguments proving the validity of the algorithm for finding the optimal PR periods for a system of arbitrary order. The reader is referred to paper [43] which contains the detailed proof. It is based on the inductive transition from the system of a lower rank to the system of a higher rank.

First we describe the procedure for the second-order system. Its subsystems of the first rank we shall simply call "subsystems" and the subsystems of the second rank we shall call "elements".

The algorithm is demonstrated by Table 12.

Step 1. Find the optimal PR periods for every separate element concerning only the ratios $\varphi(x)/x$ (see Section 1 in the table).

Step 2. Construct the basic sequence for every subsystem, taking the vector found in the previous step as an initial one (see left side of Section 2 in the table).

Step 3. For every subsystem calculate the values of $W(v)$ according to the formula given in Section 2 of the table. Find the optimal periods for each separate subsystem.

Table 12.

Algorithm for the second order system						
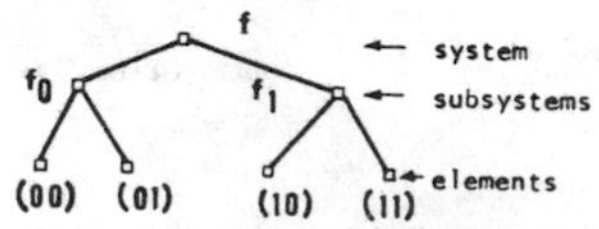					Periods available for PRs: $t_1 = \Delta,\ t_2 = 2\Delta,\ t_3 = 4\Delta,$ $t_4 = 8\Delta,\ t_5 = 16\Delta,\ t_6 = 32\Delta;$ $\Delta = 1$ Formula used for finding minimum	
16	8	8	4	Optimal periods for elements	$\min_{\{\tau_{ij}\}} \varphi_{ij}(\tau_{ij})/\tau_{ij},\ \tau_{ij} \in \{t_k\}$	1. Optimization on the elements level.
16 16 32	8* 16 32	8 8 16 32	4 8* 16 32	Basic sequences for subsystems	$\dfrac{f_i}{\min_j \tau_{ij}} + \sum_{j=0}^{1} \varphi_{ij}(\tau_{ij})/\tau_{ij}$	2. Optimization on the subsystems level.
16	8	8	8	Initial vector in the system's bas. seq.	$\dfrac{f}{\min_{i,j} \tau_{ij}} + \dfrac{f_1}{\min_j \tau_{1j}} + \dfrac{f_0}{\min_j \tau_{0j}}$ $+ \sum_{i,j=0} \varphi_{ij}(\tau_{ij})/\tau_{ij}$	3. Optimization on the whole system's level
16 16 32	8 16 32	8 16 32	8 16 32	Basic sequence for whole system		
16	16	16	16*	Optimal tuple: $V(16, 8, 8, 8) > V(16, 16, 16, 16) < V(32, 32, 32, 32)$		

Step 4. Combine together the optimal tuples found for the subsystems in one common tuple and take it as the initial tuple for the whole system (see the left side of Section 3 in the table).

Step 5. Starting with this tuple, construct the basic sequence $\{w^{(r)}\}$ for the whole system (see Section 3 in the table). Calculate the total all-system costs for this sequence according to the formula given in Section 3 of the table.

Step 6. Calculate all-system costs $W(w^{(r)})$ for vectors $w^{(r)}$ according to the formula given in Section 3 of the table.

Step 7. Find the optimal tuple $w^{(r+k)}$ satisfying the relation:

$$W(w^{(r)}) \geqslant W(w^{(r+1)}) \geqslant \cdots \geqslant W(w^{(r+k)}) < W(w^{(r+k+1)})$$

or

$$W(w^{(r)}) \geqslant W(w^{(r+1)}) \geqslant \cdots \geqslant W(w^{(r+k)}),$$

(when $r+k=K$).

The algorithm for the system of arbitrary order n might be described as follows.

The system of the order n, $n \geqslant 3$, can be presented as a union of systems having the order $(n-1)$ (see Figure 38). Assume that we have

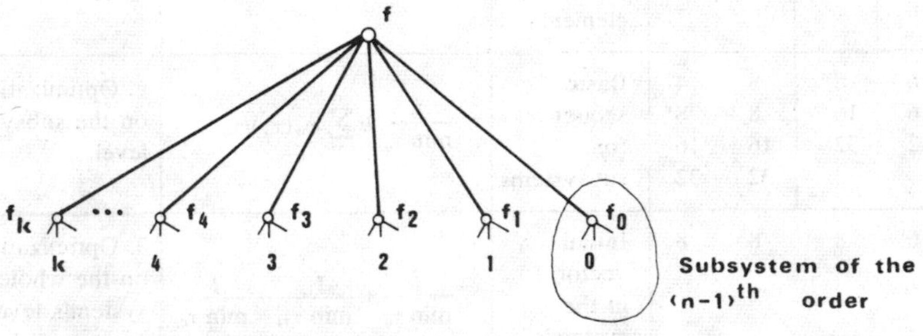

Figure 38. Explanation of the algorithm for a system of the nth order.

previously obtained the optimal tuples of PR periods for each separate subsystem of the order $(n-1)$. Denote such an optimal tuple for the ith subsystem w_i^* and compose the union of these tuples

$$w^* = (w_1^*; w_2^*; \ldots; w_k^*).$$

Now proceed with constructing the basic sequence in the following manner: the minimal members in w^* are replaced sequentially by larger ones according to the list of available PR periods (2.2). For every vector of this basic sequence a total cost has to be calculated (see general expression (2.4)). If we reveal that

$$W(w^{(m)}) \geqslant W(w^{(m+1)}) < W(w^{(m+2)})$$

or

$$W(w^{(m)}) \geqslant W(w^{(m+1)}), \qquad m+1=K,$$

then the vector $w^{(m+1)}$ defines an optimal solution.

2.5. Concluding Remarks

(1) Let us discuss the validity of the supposition which concerns itself with the convexity of $\varphi(x)$ ($\varphi''(x) > 0$).

In the case of block replacement (Chapter 2, Section 3.6) the function $\varphi(x)$ has the form:

$$\varphi(x) = M(x)c_a + c_p.$$

It can be shown that for widely used DFs, especially those having large variation coefficients, $M(x)$ will have the desired property. Thus, for the gamma distribution with parameter $k = 2$, the explicit form for $M(x)$ is:

$$M(x) = \frac{\lambda x}{2} - \frac{1}{4}(1 - e^{-2\lambda x}).$$

Hence, $\varphi''(x) > 0$.

In the model considered in Problem 5, Section 5 of Chapter 2, the function $\varphi(x)$ had the form

$$\varphi(x) = C_a \int_0^x \lambda(t)\,dt + C_p.$$

It is obvious that if the system has an increasing hazard rate ($\lambda'(t) > 0$), then $\varphi(x)$ is also convex.

In [2], the following model of system's failure was considered. In an object, damages occur according to the Poisson flow with the parameter λ. Each damage degenerates into failure independently of the others after a random time τ having DF $F(t)$. PM of the object consists of removing all damages; PRs are performed at regular time periods. It was shown in the cited paper that the average number of failures occurring between two successive PRs is equal to $\lambda \int_0^{\Delta} F(u)\,du = \Psi(\Delta)$. It is clear that here also the function $\varphi(\Delta) = C_a\Psi(\Delta) + C_p$ is convex (it is supposed that $F'(t) > 0$).

We have to note that violation of the condition $\varphi''(x) > 0$ may lead to the following consequences: the optimal vector of periodicities does not necessarily belong to the basic sequence; even if it does, it is not guaranteed that the first minimum point on the sequence $W(w^{(k)})$ ensures the optimality. Nevertheless, some numerical examples show that the above algorithm gives satisfactory or good results even if the convexity condition does not hold.

(2) Let us consider the case where we take into account restrictions on the periods of PM. Assume that for the ith element of the system, there is a restriction on the period of PM of the form

$$\tau_i < T_i.$$

Such constraints may arise, for example, if it is necessary to restrict the probability of a failure between PRs.

The formal way to include these restrictions in our model is to modify the functions $\varphi_i(\tau)$. We have to set $\varphi_i(\tau) = N$ for $\tau > T_i$, where N is a large enough value.

(3) *Possible generalizations.* The model discussed in this section might be called an "additive" one. It fits the situation when the cost of repairing a certain group of elements is a *sum* of "element" costs and additional payments caused by the involvement in the PR of systems having "higher" rank. This outline is realistic if it is necessary to pay separately for every job performed by executing a PR. Thus, we have to pay some service expense independently of what will be repaired (the "system's payment"), then the expenditures for supervising the subsystem ("subsystems' payment"), and so on up to the cost for the repair of the element itself ("elements' payment"). We suppose that such a situation is more or less typical where all kinds of repair work are highly specialized and every operation is executed by a specialized team or worker and therefore must be paid for separately.

At the same time, it may not be out of place to consider a quite different situation. Assume that simultaneously a group of elements constituting the set **I** is repaired. Then the cost of the repair (here it would be more realistic to interpret the "cost" as the time) is equal to the $\max_i \theta_i$, where θ_i is the time associated with the ith element. This picture would make sense for the repair of an important military equipment done simultaneously by several independent repair teams; the time the system spends in the repair department depends, therefore, on the longest repair time. It would be desirable to have a good working algorithm for finding the optimal PR rule for this case.

3. *Preventive Maintenance of Objects with Multidimensional Description of Their State*

3.1. General Description

In Chapter 3, several preventive maintenance models based on the observation of a one-dimensional parameter were considered. The main difficulties intrinsic in these problems were caused by the necessity to find an appropriate stochastic description for the random process. If the process describing the changes of the system's parameter permits its representation in terms of a Markov process, then the situation is more or less satisfactory.

In practice, however, a situation when the state of the system (object) is described well enough by a *one-dimensional* parameter is the exception rather than the rule. More often the correct description of the state of a technical object is given in terms of some multidimensional process

$$\boldsymbol{\eta}(t) = (\eta_1(t), \eta_2(t), \ldots, \eta_n(t)), \tag{3.1}$$

where some domain S in the n-dimensional space is a region of "acceptance". For example, the state of the braking system of an automobile under measurement on a diagnostic stand may be described by the tuple of basic parameters $(S_1, t_1, \ldots, S_4, t_4, x)$, where S_i, t_i are the braking distance and the braking period of the ith wheel and x is the output capacity of the air compressor.

There arises the question of constructing, based on observations of the vector process $\boldsymbol{\eta}(t)$, an optimal system of preventive actions. In other words, the question is about the optimal control of a multidimensional stochastic process.

There occur several situations where it turns out that the case is reduced to control of a one-dimensional process.

(a) The parameters $\eta_k(t)$, $k = 2, \ldots, n$, during the entire period of operation are connected with the parameter $\eta_1(t)$ by the relation $\eta_k(t) = f_k[\eta_1(t)] + \epsilon$ where f_k are known functions and ϵ is an observation noise. Actually, this is the case when parameters $\eta_k(t)$, $k = 2, \ldots, n$ do not contain useful information.

(b) All $\eta_i(t)$ are independent processes. Physically, this takes place if $\eta_i(t)$ depends only on a part of the system which is functionally independent of other parts.

In the example of a braking system, it is possible to assume that the pairs (S_k, t_k) are independent of each other, since they depend only on the condition of the braking device of one wheel and not on x (in a certain range of values of x). If, in this connection, it is considered that S_k is functionally connected with t_k, then the problem is simplified, and we actually have a case with five independent parameters.

In the presence of the simplifications (a) or (b), it is possible to control each parameter $\eta_i(t)$ individually. The interconnection which may arise in this case is evidenced in the execution of preventive inspection work on different parts, appropriately combined in time.

However, in the general case, when the simplifications indicated above are not assumed, the construction of optimal preventive maintenance plans becomes a very difficult problem. Aside from "ordinary" complications caused by the necessary introduction of a multidimensional "critical" region, the difficulties are sharply increased by the need for the correct probabilistic description of the multidimensional stochastic process $\boldsymbol{\eta}(t)$. They are aggravated still further in that the object for all time is under the influence of "unforeseen" effects (breakages, partial repairs, controls, and so on). This hinders observation of "pure" sample functions of $\boldsymbol{\eta}(t)$. In addition, in itself, the formal treatment of a control of a multidimensional process is very difficult.

What emerges from this situation? Let us consider the practical side of the organization of preventive inspection-repair work. For almost all devices intended for long use, there is usually established some one-dimensional characteristic according to which an opinion is formed about the object as a whole. Such a characteristic might be the total mileage (of an automobile), the accrued time in the air (of an airframe), the total operation time in hours (of a machine), the efficiency of a production cycle, the cost of one overhaul (see, for example, [45]), repair costing required to maintain a vehicle [3], and so on.

There thus arises a natural desire to replace the vector process $\boldsymbol{\eta}(t)$ by a scalar process, by constructing some scalar function $r(t)$ of the components $\eta_i(t)$, $i, \ldots, n$, and to exploit the new parameter $r(t)$ as a basis for making decisions about repairs, inspections, etc.

3.2. The Best Scalarization

One of the most essential concepts of scalarization is that the scalar function

$$r(t) = f(\eta_1(t), \eta_2(t), \ldots, \eta_n(t)) \tag{3.2}$$

must have the best discrimination properties with respect to the non-failed and failed objects. Hence, there follows the possibility to use some of the ideas of discriminant analysis [1, Chapter 6], [71, Chapter 8]. From the practical point of view, it is very convenient to use Fisher's linear discriminant function [33]. Let us consider how this function is constructed.

Let us assume that the sample of objects can be conventionally divided into two groups: A (the nonfailed or "good") and B (the failed or "unfit"). It must be specified that such a subdivision cannot be done always easily. Medical men usually can distinguish very well between a healthy person and a sick one, and, moreover, classify the patient within a specific group according to his illness. In technology, such a classification is sometimes very difficult. Indeed, between the states "brand-new object" and "entirely useless object" there are numerous intermediate states difficult to distinguish. In practical life, the decision is usually made on the basis of intuitive reasoning plus a background of experience and the application of common sense, etc.

In our case, we will need to distinguish between two "extreme" categories of objects: "very good" and "very bad". Thus, we begin by assuming that there is an expert who can classify all objects under consideration into two groups, A and B. Let N_A and N_B be the numbers of elements in these two groups, which are characterized by samples

$$\{\boldsymbol{x}_1^A, \ldots, \boldsymbol{x}_{N_A}^A\} \quad \text{and} \quad \{\boldsymbol{x}_1^B, \ldots, \boldsymbol{x}_{N_B}^B\}, \tag{3.3}$$

where every element of the sample $\boldsymbol{x}_j^z$, $z = A, B$; $j = 1, N_z$, is represented by a n-tuple

$$\boldsymbol{x}_j^z = \{x_{j1}^z, \ldots, x_{jn}^z\}. \tag{3.4}$$

Fisher's idea of discrimination is based on the replacement of the vector value $\boldsymbol{x}_j^z$ by a scalar

$$\sum_{i=1}^{n} l_i x_{ji}^z, \tag{3.5}$$

where the coefficients of linear transformation l_i are selected in an optimal way. Geometrically, the samples A and B are two clusters of points in the n-dimensional space. Let $\hat{\boldsymbol{\mu}}_A$, $\hat{\boldsymbol{\mu}}_B$ be the n-dimensional vector of mean values and $\mathbf{S}_A$ and $\mathbf{S}_B$ be the matrices of sum of squares and products for samples A and B respectively:

$$\hat{\mu}_A^{(i)} = \sum_{m=1}^{N_A} x_{mi}^A / N_A, \qquad \hat{\mu}_B^{(i)} = \sum_{m=1}^{N_B} x_{mi}^B / N_B;$$

$$\hat{\boldsymbol{\mu}}_z = (\hat{\mu}_z^{(1)}, \ldots, \hat{\mu}_z^{(n)}), \qquad z = A, B.$$

$$\mathbf{S}_A = \| v_{ij}^A \|, \quad i, j = 1, \ldots, n; \qquad \mathbf{S}_B = \| v_{ij}^B \|. \tag{3.6}$$

$$v_{ij}^z = \sum_{m=1}^{N_z} (x_{mi}^z - \hat{\mu}_z^{(i)})(x_{mj}^z - \hat{\mu}_z^{(j)}), \qquad z = A, B.$$

Linear transformation signifies replacement of each observation $\boldsymbol{x}_j^z$ by the scalar $\Sigma_{i=1}^n l_i x_{ji}^z$, which is geometrically equivalent to the projection of each cluster onto some line, the direction of which is given by the vector $\boldsymbol{l} = (l_1, \ldots, l_n)$. It is desirable to choose the direction of the line so that under projection, the populations A and B are differentiated in the best possible way.

Figure 39 shows two clusters A and B in a two-dimensional space. It is easy to check that the direction $\boldsymbol{l}^{(1)}$ discriminates between the clusters better than the direction $\boldsymbol{l}^{(2)}$.

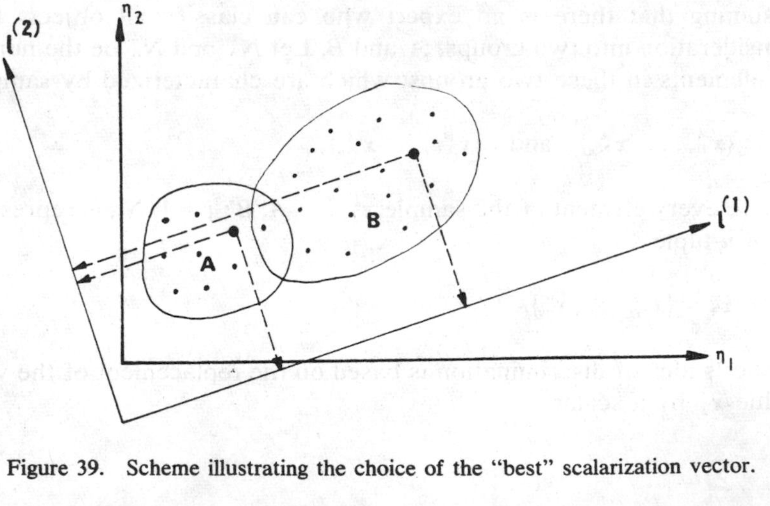

Figure 39. Scheme illustrating the choice of the "best" scalarization vector.

Fischer proposed to choose the direction $\boldsymbol{l}$ in such a way that the ratio of squares of the difference of the mean values on the line $\boldsymbol{l}$ over the sum of variations would be maximal.

Let every observation be replaced by the projection

$$y_j^z = l_1 x_{j1}^z + \cdots + l_n x_{jn}^z = \boldsymbol{l}\, \boldsymbol{x}_j^z, \tag{3.7}$$

where l is a row-vector, $z = A, B$, and x_j^z is a column-vector. The mean value in the sample z is

$$\hat{y}^z = \sum_{j=1}^{N_z} \sum_{k=1}^{n} x_{jk}^z l_k / N_z = l\hat{\mu}_z. \tag{3.8}$$

($\hat{\mu}_z$ is a column-vector).

The variance of the projection of both samples is given by

$$\mathbf{S} = \sum_{z=A,B} \sum_{j=1}^{N_z} (y_j^z - \hat{y}_z)^2 = \sum_{z=A,B} \sum_{j=1}^{N_z} (l x_j^z - l\hat{\mu}_z)^2$$

$$= l\left\{ \sum_{z=A,B} \sum_{j=1}^{N_z} (x_j^z - \hat{\mu}_z)(x_j^z - \hat{\mu}_z)^T \right\} l^T = l(\mathbf{S}_A + \mathbf{S}_B)l^T. \tag{3.9}$$

(Notation T means transposition.) Therefore, it is a matter of finding a vector l in such a way as to maximize the value

$$\varphi = \frac{\{l(\hat{\mu}_A - \hat{\mu}_B)\}^2}{l(\mathbf{S}_A + \mathbf{S}_B)l^T}. \tag{3.10}$$

It is clear that the components of the vector l are determined only up to a constant multiplier. Let $(\mathbf{S}_A + \mathbf{S}_B)$ be a nonsingular matrix. It is a well-known fact from linear algebra (see, for example [1, Appendix]) that there is a nonsingular matrix $\mathbf{C}$ such that

$$\mathbf{C}(\mathbf{S}_A + \mathbf{S}_B)^{-1}\mathbf{C}^T = \mathbf{I}. \tag{3.11}$$

Then set

$$\mathbf{Y} = \mathbf{C}l^T.$$

Now

$$\begin{aligned} l(\mathbf{S}_A + \mathbf{S}_B)l^T &= \mathbf{Y}^T\mathbf{C}^{-1^T}(\mathbf{S}_A + \mathbf{S}_B)\mathbf{C}^{-1}\mathbf{Y} = \mathbf{Y}^T\mathbf{Y} = (\mathbf{Y}, \mathbf{Y}), \\ \varphi &= \{\mathbf{Y}^T, \mathbf{C}^{-1^T}(\hat{\mu}_A - \hat{\mu}_B)\}^2/(\mathbf{Y}, \mathbf{Y}), \end{aligned} \tag{3.12}$$

where in the numerator of φ there appears the scalar product of the corresponding vectors. Clearly, it suffices to limit attention to those $\mathbf{Y}$ which satisfy $(\mathbf{Y}, \mathbf{Y}) = 1$. Then, obviously, the maximum value of φ in (3.12) will be achieved on a vector $\mathbf{Y}$ which is collinear with $\mathbf{C}^{-1^T}(\hat{\mu}_A - \hat{\mu}_B)$, i.e. up to a constant multiplier, $\mathbf{Y} = \mathbf{C}^{-1^T}(\hat{\mu}_A - \hat{\mu}_B) = \mathbf{C}l^T$, from which

$$l^T = \mathbf{C}^{-1}\mathbf{C}^{-1^T}(\hat{\mu}_A - \hat{\mu}_B) = (\mathbf{S}_A + \mathbf{S}_B)^{-1}(\hat{\mu}_A - \hat{\mu}_B). \tag{3.13}$$

Thus, to obtain the optimal vector, it is necessary to calculate the inverse matrix $(\mathbf{S}_A + \mathbf{S}_B)^{-1}$ and multiply it by the vector $(\hat{\mu}_A - \hat{\mu}_B)$.

It is interesting to compare this result with another approach used in mathematical statistics for the classification of objects into two groups. Let the probabilistic properties of populations A and B be described by their density functions $p_A(\mathbf{x})$ and $p_B(\mathbf{x})$ respectively. According to the theorem of Neyman-Pearson, the best inference about the membership relation of a given object $\mathbf{x}_0$ to populations A or B may be formed on the basis of the likelihood ratio

$$q_{AB}(\mathbf{x}_0) = p_A(\mathbf{x}_0)/p_B(\mathbf{x}_0)$$

(or any monotone function $r(q)$ of it).

If p_A and p_B are multidimensional normal densities with known mean vectors $\boldsymbol{\mu}_A$ and $\boldsymbol{\mu}_B$ and the same (nonsingular) covariance matrices $\mathbf{V}$, then by using standard manipulations, we obtain

$$r_{AB}(\mathbf{x}_0) = \ln q_{AB}(\mathbf{x}_0) = \mathbf{x}_0^T\mathbf{V}^{-1}(\boldsymbol{\mu}_A - \boldsymbol{\mu}_B) - \tfrac{1}{2}(\boldsymbol{\mu}_A + \boldsymbol{\mu}_B)^T\mathbf{V}^{-1}(\boldsymbol{\mu}_A - \boldsymbol{\mu}_B).$$

In this formula, the second term is a constant value which does not depend on actual observations and the first term is nothing but a scalar product of the vector $\mathbf{x}_0$ by the vector $\mathbf{V}^{-1}(\boldsymbol{\mu}_A - \boldsymbol{\mu}_B)$. This demonstrates the analogy between the Fisher and the Neyman-Pearson approaches.

From the practical point of view, Fisher's discriminant function is preferable because in computing it, no assumption about the multidimensional densities must be made.

3.3. How to Organize Preventive Maintenance of an Object with a Multidimensional Description of Its State

At this point, we intend to describe the procedure and processing data for preventive maintenance actions based on a multidimensional parameter. Next, we shall present a corresponding numerical example. To make the matter clearer we shall divide the procedure into several steps.

Step 1. Two collections of "brand new" and "entirely unfit" objects are selected. It is desirable that the first group consist of good-functioning units which have been in operation for a relatively short time. The second should contain units which were in operation for a longer time, and according to the estimation of an expert, can be considered as failed. The actual values of the parameters have to be measured for every object, and samples of type (3.3) must be formed. We draw attention to the fact that it is not expedient to restrict ourselves *a priori* to the number of measured parameters. The more parameters it is possible to check, the better will be the results.

Step 2. The vectors $\hat{\boldsymbol{\mu}}_A$, $\hat{\boldsymbol{\mu}}_B$, the matrices $\mathbf{S}_A$, $\mathbf{S}_B$, and $(\mathbf{S}_A + \mathbf{S}_B)^{-1}$ are calculated, and finally also the vector $\boldsymbol{l}$ which is defined in (3.13). By virtue of vector $\boldsymbol{l}$, every n-tuple of observations is reduced to one value $\{y_j^A\}_{j=1}^{N_A}$ or $\{y_j^B\}_{j=1}^{N_B}$ (see (3.7)). Now these two one-dimensional samples must be investigated. We recommend plotting histograms for these two samples. A decision must be made if the discrimination is satisfactory or not. In an extreme situation such as that shown in Figure 40, the conclusion is obvious. Thus, if something similar to Figure 40 (left)

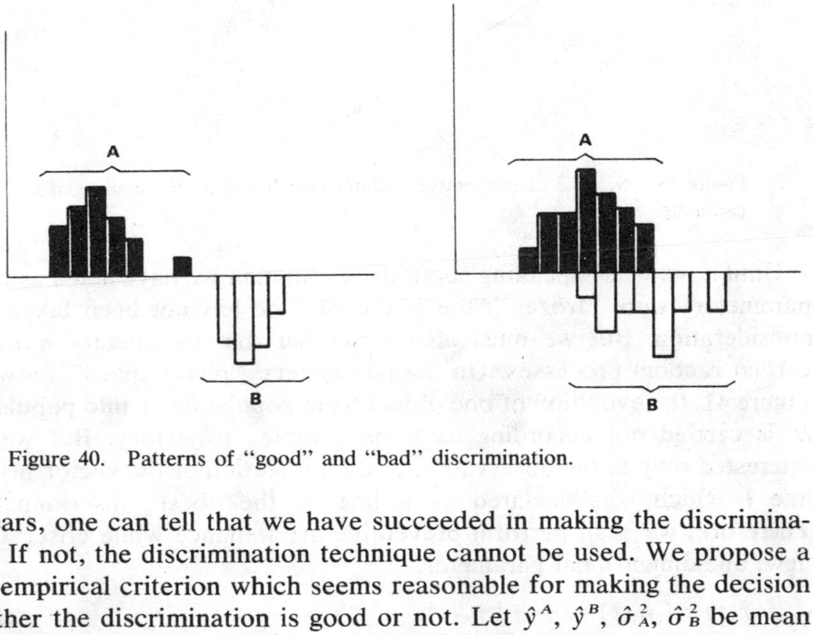

Figure 40. Patterns of "good" and "bad" discrimination.

appears, one can tell that we have succeeded in making the discrimination. If not, the discrimination technique cannot be used. We propose a half-empirical criterion which seems reasonable for making the decision whether the discrimination is good or not. Let $\hat{y}^A$, $\hat{y}^B$, $\hat{\sigma}_A^2$, $\hat{\sigma}_B^2$ be mean values and variances for the obtained one-dimensional samples. We assume the discrimination to be good if

$$|\hat{y}^A - \hat{y}^B| > 2.5(\hat{\sigma}_A + \hat{\sigma}_B). \tag{3.14}$$

Figure 41 shows one possible way to organize preventive maintenance using the multidimensional state parameter. Consider a certain object having at the time t_0 the vector of parameters

$$\boldsymbol{\eta}(t_0) = (\eta_1(t_0), \ldots, \eta_n(t_0)). \tag{3.15}$$

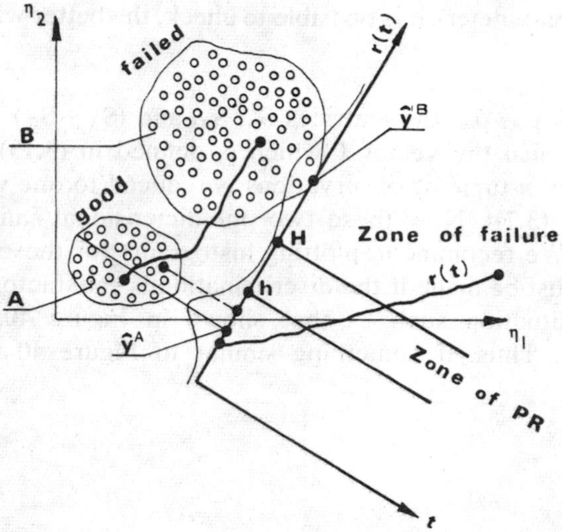

Figure 41. Scheme of preventive maintenance based on the control of the scalar parameter $r(t)$.

Until now, while speaking about discrimination we have acted as if the parameters were "frozen"; the factor of time has not been taken into consideration. But we must also remember that parameters $\eta_i(t)$ are certain random processes. On the parameters' "phase space" shown in Figure 41, the evolution of one object from population A into population B is carried out according to some complex trajectory. But we are interested only in the observation of the projection of the vector $\boldsymbol{\eta}(t)$ on line l which was declared as a line of the "best" discrimination. Therefore, we shall perform preventive maintenance while observing a new, one-dimensional parameter.

$$r(t) = l_1\eta_1(t) + \cdots + l_n\eta_n(t). \tag{3.16}$$

We are justified in doing this because in the area where this parameter changes, we have clearly expressed zones corresponding to "good" and "failed" objects.

Evidently, it is expedient to introduce two levels, H and h, so that the situation $r > H$ might be considered as a failure when an ER must be done, and the state $h \leq r \leq H$ as a marginal situation when a PR must be made. Thus, we are in the familiar one-dimensional situation which allows standard preventive maintenance routine.

It remains to describe how we can obtain data concerning the statistical properties of the random process $r(t)$.

Step 3. This consists of observing a group of operating objects with the purpose of obtaining the statistical description of $r(t)$.

We advise the reader to study the statistical properties of $r(t)$ by observing individual sample functions of this process, each of which can be obtained on the basis of periodic observations and measurements of a specific operating object. Let us recall that in real circumstances, the "time" can be the actual operating time, the run measured in kilometers (for an automobile, for example), the total number of switchings (for electrical equipment), the total number of landings and takeoffs (for aircraft) and so on. Thus, if the measurements executed at times $t_1, t_2, \ldots, t_k$ for one object gave the results $\boldsymbol{\eta}(t_1), \boldsymbol{\eta}(t_2), \ldots, \boldsymbol{\eta}(t_k)$, then we have a discrete sample function of $r(t)$ having the value $r(t_i) = \Sigma_{j=1}^{n} l_j \eta_j(t_i)$ at time t_i.

The statistical inferences about the process $r(t)$ poses a difficult question. There are some reasons to presume that $r(t)$ would be a process with independent increments or a Markovian-type process. $r(t)$ is a sum of several components, and in many cases these may have relatively weak statistical dependence (or correlation). The operating object is always influenced by changing outside factors such as changes of the load or of the service conditions, partial repairs, etc. As a result of all these reasons it may happen that the increments of the process $r(t)$ on the nonoverlapping time intervals will be stochastically independent. In [41] a procedure is described for carrying out the statistical analysis of sample functions corresponding to accumulation of damages or wear deterioration. This technique should be applied to sample curves of the observed process $r(t)$.

3.4. Preventive Maintenance of Hydraulic Pumps Using Scalarization

At this point we demonstrate all three steps given in the previous section in reference to hydraulic axial pumps used in aircraft. A model of preventive maintenance will also be given and numerical data be presented.

According to available statistical data, about a quarter of all failures in aircraft are caused by malfunctions in their hydraulic systems; 30% of these appear in the most complicated and responsible devices such as hydraulic pumps, hydraulic actuators, etc. That is why particular attention has been paid to the study of the reliability of aircraft hydraulic pumps and especially to their diagnostics.

Data discussed below are related to axial hydraulic pumps. Their failures are caused mainly by deterioration of pumping devices, destruction of axial bearings and wear and tear of packings. Observations have shown that after a period of 3000 hours of operation (which was stated as a critical age) a significant percentage of the pumps were in a good state with respect to their output characteristics. This means that it is more expedient to make replacements not based on the time which has elapsed since the start of operation, but by taking into consideration the data received from measurement of parameters about the "inside" state of the operating object.

The most effective diagnostic methods which do not involve dismantling of the operating object are based on the measuring of oil pressure by means of special sensors. In the presence of the end play of a hydraulic piston (the main cause of malfunction occurring in this sort of hydraulic pumps), a backward flow of oil comes from the blower line to the working chamber. This causes a step-up of pressure vibration in the hydraulic system. The important index which evidences the actual state of the pump is the coefficient

$$\eta^* = \Delta P/P,$$

where $\Delta P = P_{max} - P_{min}$ is the pressure oscillation amplitude; P – the average pressure in the line. It was ascertained that the value η^* is highly correlated with the total amount of the plungers' end play. Another efficient method of diagnostics is based on the fact that the increase in the deterioration of the pump's elements (plungers, couples, bearings, packings, etc.) causes the changes in the vibration spectrogram. These changes correspond to the main plunger rotation frequency multiplied by the number of plungers and to the bearings' rotation frequency. Data received by means of piezocrystal sensors show an increase in the value of the average square of acceleration amplitude a. Similar is the situation with the spectres of vibration shifting δ. An index of great diagnostic importance is also the pump's output efficiency η_0.

A sample of 33 pumps which have different operation times was examined. Every unit was tested and eight parameters were measured: η^*, η_0, a and δ for three different frequencies. Later on, every pump was dismantled, carefully tested and the final decision about its state was made by an expert. Nine pumps were recognized as failed.

Table 13 represents the results obtained by performing Step 2. The first part of this table shows data concerning only two parameters which seem intuitively to be the most important. However, it may be concluded that the discrimination is not good. Thus, the ratio

$$|\hat{y}^A - \hat{y}^B|/(\tilde{\sigma}_A + \tilde{\sigma}_B) = 1.08.$$

When five parameters were involved in the scalarization, the results were better:

$$|\hat{y}^A - \hat{y}^B|/(\tilde{\sigma}_A + \tilde{\sigma}_B) = 1.52.$$

The scalarization based on all measured parameters gave quite good results as is seen from the lower section of Table 13. The discrimination ratio is

$$|\hat{y}^A - \hat{y}^B|/(\tilde{\sigma}_A + \tilde{\sigma}_B) = |339.6 - 311.9|/(4.99 + 6.00) = 2.52.$$

The following is proposed on the basis of these data: if $r > 333$, the pump is considered as "good"; if $r < 323$, then the pump is recognized as having failed; the intermediate situation, $323 \leq r \leq 333$, is considered as marginal and preventive replacement must be done.

Table 14 represents data received from the observation of 17 pumps operating under "field" circumstances. Each pump was tested several times at different operating times since it began to function, and all the above-mentioned parameters were measured. For each unit a pointwise sample function of $r(t)$ was obtained. For the sake of simplicity, the scale was changed and set $r^*(t) = (362 - r(t))\,0.4$.

Processing of these data, and, in particular, the study of the slope of $r^*(t)$, permit the following conclusions:

(i) there is no correlation between the slopes of $r^*(t)$ on the nonoverlapping time intervals ($\rho = -0.077$);
(ii) there is no correlation between the slope and the flight time (operation time) $\rho = 0.008$;
(iii) there is only a slight correlation between the slope of $r^*(t)$ and the actual value of $r^*(t)$: $\rho = 0.31$ by $\sigma_\rho \simeq (1 - \rho^2)/\sqrt{n} = 0.13$.

All this provides a reason for accepting the hypothesis that $r^*(t)$ is a process with independent increments. If we observe the states of that process in discrete time instants, we obtain a time- and space-homogeneous Markov chain.

It is easy to calculate from the data of Table 14 that the increments of

Table 13.

No.	Parameters involved in scalarization	Coeff. in the discrim. function	Parameters of onedim. samples: A good	Parameters of onedim. samples: B failed	Histograms of one-dimensional samples (○ good, ● failed)
1	η^* η_0	−20.5 112.8	$\hat{y}^A = 102.0$ $\hat{\sigma}_A = 1.48$	$\hat{y}^B = 95.9$ $\hat{\sigma}_B = 4.16$	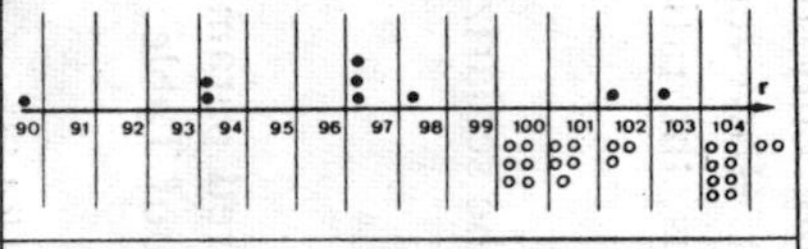
2	η^* a_1 δ_1 a_2 δ_2	−28.4 −0.063 26.8 −0.043 38.8	$\hat{y}^A = 53.6$ $\hat{\sigma}_A = 2.07$	$\hat{y}^B = 41.7$ $\hat{\sigma}_B = 5.79$	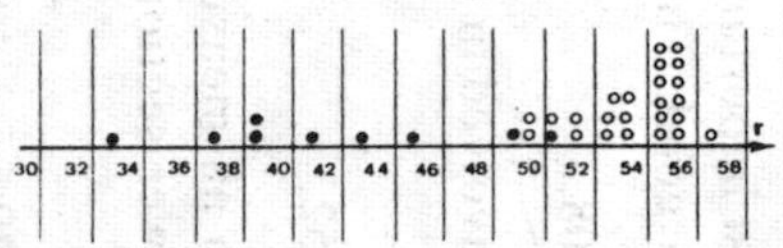
3	η^* a_1 δ_1 a_2 δ_2 a_3 δ_3 η_0	−65.8 −0.196 3.48 0.197 28.6 −0.038 131.5 215.7	$\hat{y}^A = 339.6$ $\hat{\sigma}_A = 4.99$	$\hat{y}^B = 311.9$ $\hat{\sigma}_B = 6.00$	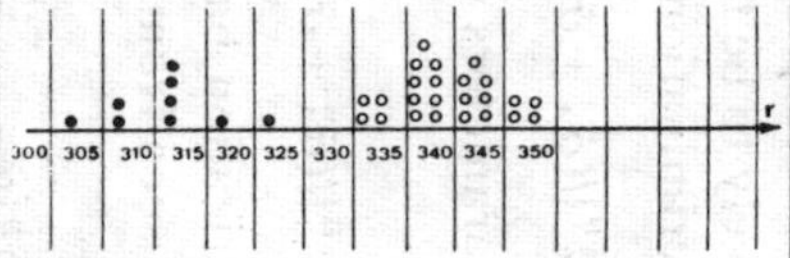

Table 14.

No.	Ordinal number of the pump	Operation time (hours)	$r^*(t)$	No.	Ordinal number of the pump	Operation time (hours)	$r^*(t)$
1	1	2290	12.43	33	9	620	9.01
2	1	2500	11.96	34	10	1000	7.44
3	1	2790	15.78	35	10	1200	11.33
4	1	2950	18.03	36	10	1790	10.34
5	1	3130	17.04	37	11	1560	11.39
6	1	3240	17.04	38	11	1920	15.10
7	2	307	9.83	39	11	2120	16.58
8	2	900	10.18	40	11	2425	11.34
9	3	324	12.23	41	11	2720	11.92
10	3	930	10.17	42	12	2330	14.29
11	4	630	7.90	43	12	2590	13.75
12	4	840	10.27	44	12	2780	14.19
13	4	1040	10.29	45	12	3000	12.87
14	4	1230	10.29	46	12	3200	15.43
15	5	2630	7.89	47	12	3500	16.33
16	5	2891	7.89	48	12	3800	15.32
17	5	3220	9.93	49	13	600	6.81
18	5	3520	12.56	50	13	800	6.81
19	6	1030	8.39	51	13	1030	7.27
20	6	1230	8.03	52	13	1900	7.73
21	6	1400	9.51	53	14	1960	7.34
22	6	1600	13.51	54	14	2170	9.18
23	6	2400	14.64	55	14	2380	8.44
24	7	1830	8.58	56	14	2570	7.38
25	7	2030	10.05	57	14	2850	9.17
26	7	2300	12.39	58	14	3160	10.05
27	7	2600	14.70	59	15	217	7.71
28	7	3050	14.92	60	15	420	8.06
29	8	603	8.74	61	15	540	8.00
30	8	1270	9.64	62	16	2630	9.01
31	9	220	8.40	63	16	2780	9.53
32	9	420	8.03	64	17	60	6.73
				65	17	240	7.73

the process $r^*(t)$ on the 200-hour interval are distributed approximately normally and that the parameters are:

$$\Delta r^*_{200} = r^*(200) - r^*(0) \sim \mathfrak{N}(\hat{r}_{200} = 0.72; \hat{\sigma}_{200} = 1.3)$$

Henceforth, the period $\Delta = 200$ hours will be considered as a unit-time interval. To simplify the calculations, we introduce several discrete levels for the process $r^*(t)$. Thus, it is assumed that $r^*(t)$ has 25 levels – $E_1, \ldots, E_{25}$. Level E_1 corresponds to $r^*(t)=4.0$ and each succeeding level means an increase in $r^*(t)$ of 0.5. The region $E_{16}, \ldots, E_{24}$, is considered as marginal; if anyone of these states is discovered by inspection, the pump is replaced by a new one. States E_{25} and higher are failure states; states $E_1, \ldots, E_{15}$ are considered as "good".

The transition probabilities for the Markov chain arising when the process $r^*(t)$ is observed periodically were evaluated as follows:

$$p_{ij} = \boldsymbol{P}\{r_j^* - r_i^* - 0.25 < \xi \leq r_j^* - r_i^* + 0.25\},$$

where ξ is a random increment of the process $r^*(t)$ during fixed time interval Δ, and E_j corresponds to $r^*(t) = r_j^*$.

For the boundary states E_1 and E_{25}, obvious changes must be done in the calculations of the transition probabilities:

$$p_{k,1} = \boldsymbol{P}\{\text{to go from } E_k \text{ into the region } r(t) < 4\},$$
$$p_{k,25} = \boldsymbol{P}\{\text{to go from } E_k \text{ into the region } r(t) > 16\}.$$

Now assume that the repair group can rapidly test the pump and measure the actual values of its parameters, as a result of which the value of scalar parameter $r^*(t)$ will be obtained.

It is assumed also that the repair team discovers the failure immediately after it occurs, for example, during the post-flight checkup. In that case, a replacement called ER will be performed.

Every replacement means, formally, a transferring from state E_k into a lower state. Since all the new pumps do not have the same level of parameter $r_0^*(t)$, it will be assumed that the repair of type $E_k \to E_i$ is made with probability q_{ki}. These probabilities were obtained by the supposition that new pumps, according to the observations, have an initial distribution of $r^*(t) \sim \mathfrak{N}(\hat{r}_0=8,\ \hat{\sigma}_0=2.58)$. Therefore, the transition probabilities from the state E_{25} were calculated according to the formula:

$$p_{25,s} = \sum_{j=1}^{25} \tilde{p}^{(j)} p_{js},$$

where $\tilde{p}^{(j)}$ is the probability of installing a new pump with initial level of $r_0^*(t) = E_j$: $\tilde{p}^{(j)} = \boldsymbol{P}\{r_j^* - 0.25 < r_0^*(t) \leq r_j^* + 0.25\}$.

The following service rule is accepted. In states E_i, $i = 1, \ldots, 15$, one of two alternatives can be selected: to inspect the pump after $\Delta = 200$ hours or after $2\Delta = 400$ hours. If the marginal state is discovered, the pump is replaced by a new one and the two mentioned alternatives exist. In the case of failure, which will be discovered during the post-flight service, the pump is also replaced by a new one and a new inspection time is selected.

To put into operation all SMP technique, there remains only to determine the average transition times and costs for ER and PR.

The transition from state E_i to state E_j, $j \neq 25$, lasts Δ or 2Δ depending on the selected inspection period. Values $\nu_{i,25}$ were calculated based on the supposition that the process $r^*(t)$ has a linear sample function and that the instant of failure coincides with the moment when it crosses the level E_{25}. The preventive maintenance policy considered here may be described as follows:

$$\boldsymbol{\sigma}=\{E_{\varphi(i)}, T_{\varphi(i)}; i=1, \ldots, n\}.$$

The following costs are introduced: c_s – cost for the inspection if no replacement has been made; c_p – cost for the inspection when a PR was made; c_a – cost for the failure and ER. Therefore, we are in a situation similar to that described in Section 7 of Chapter 3; our objective is to find a strategy $\boldsymbol{\sigma}^*$ minimizing the average cost rate per unit time. The problem is solved by means of Howard's algorithm. Since there are some difficulties in determining the exact values of costs associated with repair actions, all calculations were performed for several variants:

Variant 1: $c_s=1$, $c_p=10$, $c_a=100$.

Variant 2: $c_s=1$, $c_p=10$, $c_a=180$.

Variant 3: $c_s=1$, $c_p=10$, $c_a=50$.

The following table lists the optimal CAs for states E_i for all these variants (see Table, p. 244).

The reader's attention must be drawn to an interesting circumstance:

The optimal strategy is not sensitive to relatively big changes in the costs. Thus, for states $E_1 \div E_9$, $E_{13} \div E_{17}$, $E_{19} \div E_{25}$, the optimal CAs are the same. The optimal strategy has an intuitively clear form: for "good" pumps, i.e. for low levels, the inspection period is allowed to be large – 400 hours; for marginal states – smaller – 200 hours.

Now, assume that we have refused to do PR and perform only ER when failure occurs. Then for the data of Variant 3, we shall obtain the cost rate $g = 0.0164$ hours^{-1}, that is, 60% higher than in the above optimal case.

We conclude this section with the following comment. In every case when we do not have at our disposal exact information about costs, it is

State	CA VAR1	VAR2	VAR3	Notations
E_1	0; 2	0; 2	0; 2	"0" – do nothing
⋮	0; 2	0; 2	0; 2	"*" – replacement
E_9	0; 2	0; 2	0; 2	"1" – inspection after 200 hrs.
E_{10}	0; 2	0; 1	0; 2	"2" – inspection after 400 hrs.
E_{11}	0; 2	0; 1	0; 2	
E_{12}	0; 1	0; 1	0; 2	
E_{13}	0; 1	0; 1	0; 1	
E_{14}	0; 1	0; 1	0; 1	
⋮	0; 1	0; 1	0; 1	
E_{17}	0; 1	0; 1	0; 1	
E_{18}	*; 1	*; 1	0; 1	
E_{19}	*; 1	*; 1	*; 1	
⋮	*; 1	*; 1	*; 1	
E_{25}	*; 1	*; 1	*; 1	
Optimal cost rate	0.0109 hrs^{-1}	0.0113 hrs^{-1}	0.0102 hrs^{-1}	

desirable to calculate several variants as shown. This enables us to obtain data about stable domains in the strategy space and therefore makes possible the best choice even under uncertain circumstances.

APPENDIX

1. *Definition of the Semi-Markov Process ([69] and [70])*

Consider a two-dimensional random process $\{(I_m, X_m), m = 0, 1, \ldots\}$ defined on the probabilistic space $\{\mathfrak{U}, \mathfrak{B}, \mathfrak{F}\}$. The values of the component I_m belong to a finite set $E = \{E_1, \ldots, E_n\}$, $0 \leq X_m < \infty$. The following joint conditional distributions are given:

$$P\{I_m = E_k, X_m \leq x | I_0, I_1, X_1, \ldots, I_{m-1}, X_{m-1}\}$$
$$= P\{I_m = E_k, X_m \leq x | I_{m-1}\} = Q_{I_{m-1},k}(x), \tag{A.1}$$

where $0 \leq x < \infty$, $k = 1, \ldots, n$; $Q_{ij}(x)$ satisfy the following conditions:

(a) $Q_{ij}(t) = 0$ for $t < 0$;

(b) $\lim_{t\to\infty} \sum_{j=1}^{n} Q_{ij}(t) = 1, \qquad$ for $i = 1, 2, \ldots, n$; (A.2)

(c) If $Q_{ij} \not\equiv 0$, then $0 < \int_0^\infty x \, dQ_{ij}(x) < \infty$.

The initial conditions are: $X_0 = 0$, $I_0 = E_{i_0}$. Let $S_m = \Sigma_{i=0}^{m} X_i$, and define $N(t)$ as follows:

$$N(t) = \sup\{m \geq 0 : S_m \leq t\}. \tag{A.3}$$

The random process

$$\xi_t = \{I_{N(t)}, t \geq 0\} \tag{A.4}$$

is called semi-Markovian.

From (A.1) it follows that $\{I_m\}$ is a Markov chain imbedded in the process ξ_t. This chain has the transition probabilities

$$P\{I_{m+1} = E_k | I_m = E_i\} = Q_{ik}(\infty) = p_{ik}. \tag{A.5}$$

The conditional distribution of random variable X can be obtained from (A.1) and (A.2):

$$P\{X_m \leq t | I_{m-1} = E_i, I_m = E_j\}$$
$$= \frac{P\{I_m = E_j, X_m \leq t | I_{m-1} = E_i\}}{P\{I_m = E_j | I_{m-1} = E_i\}} = \frac{Q_{ij}(t)}{Q_{ij}(\infty)} = F_{ij}(t). \tag{A.6}$$

(it is assumed that $Q_{ij}(\infty) \neq 0$). It is clear that $F_{ij}(t)$ is a DF.

The SMP described in Chapter 1, Section 3 is related to the above process in the following manner. I_m is the state of the process ξ_t on the mth step; X_m – is the one-step holding time ("sitting" time) in the state I_{m-1}; $t_1 = X_1$, $t_2 = X_1 + X_2, \ldots, t_r = \Sigma_1^r X_i$ – moments of transitions. Let t_k and t_{k+1} be two adjacent transition moments and $\xi_{t_k} = E_i$. Then p_{ij} is the probability that the next state will be E_j; $Q_{ij}(t)$ is the joint probability that at time t_{k+1} the process will enter state E_j and that the interval $X_k = t_{k+1} - t_k$ will be less than t. $F_{ij}(t)$ is the DF of the transition time.

2. Probabilistic Interpretation of the Expression

$$g = \sum_{i\in J} \pi_i \nu_i \Big/ \sum_{i=1}^{n} \pi_i \nu_i \text{ – (3.46), Chapter 1.}$$

Let $P_{ij}(t)$ denote the probability that $\xi_t = E_j$ calculated under condition that $\xi_0 = E_i$; l_{ij} is the expected first-passage time from state E_i into state E_j; l_{jj} denotes the expected return time for state E_i. Let ν_i denote the ME of one-step transition time for state E_i. The following formula is obvious:

$$\nu_i = \sum_{j=1}^{n} p_{ij}\mu_{ij} = \sum_{j=1}^{n} \left\{ \int_0^\infty \tau \, dF_{ij}(\tau) \right\} P_{ij},$$

where $\mu_{ij} = M\{\tau(i, j)\}$ is the ME of one-step transition time calculated under the condition that the initial state was E_i and the transition went to E_j.

Theorem 1 (Smith [79]). *If ξ_t is a SMP with finite number of states and ergodic embedded Markov chain, $\nu_i < \infty$ for all i, then*

$$\lim_{t\to\infty} P_{ij}(t) = \nu_j / l_{jj}. \tag{A.7}$$

Theorem 2 (Barlow [4]). *If all premises of Theorem 1 hold true, then*

$$l_{jj} = \frac{1}{\pi_j} \sum_{i=1}^{n} \pi_i \nu_i, \tag{A.8}$$

where $\{\pi_i,\ i = 1, \ldots, n\}$ are the stationary limiting probabilities of the Markov chain with transition probabilities (A.5).

Corollary. *If $E_J = \{E_i, i \in J\}$ is a certain set of states, then*

$$\lim_{t\to\infty} \boldsymbol{P}\{\xi_t \in \boldsymbol{E}_J \mid \xi_0 = E_i\} = \sum_{i\in J} \pi_i \nu_i \Big/ \sum_{i=1}^{n} \pi_i \nu_i. \tag{A.9}$$

Proof. Substitute (A.8) into (A.7) and add for $i \in \boldsymbol{J}$. #

Remark 1. The formula (A.8) can be used also in the case when the Markov chain has one closed set of states but is a periodic one. Let $\boldsymbol{P} = \|p_{ij}\|$ be the matrix of this chain. It can be shown that the set of equations

$$\begin{aligned} \boldsymbol{\pi} &= \boldsymbol{\pi P}, \\ \sum_{1}^{n} \pi_i &= 1. \end{aligned} \tag{A.10}$$

has a unique, positive solution. Repeating all arguments of Barlow's theorem and using the solution of the system (A.10) instead of the vector of limiting probabilities, we obtain the formula (A.8).

Remark 2. Relation (3.43) from Chapter 1 is valid under conditions which always hold true for practical situations. If

(a) for a fixed strategy $\boldsymbol{\sigma}$ all DFs $F_{ij}(t)$ have finite first and second moments;
(b) among the DFs $F_{ij}(t)$ there is one nonlattice function;
(c) all states $E_1, E_2, \ldots, E_n$ are communicating,

then the asymptotical behaviour of the total expected costs is determined by formula (3.43) of Chapter 1.

The condition (c) can be replaced by one permitting the chain to have only one closed subset of states and a subset of transient states.

If all DFs are defined on the same lattice having step h, then the asymptotical behaviour of the expected costs is given by the formula (see [66]):

$$v_i(kh) \underset{k\to\infty}{=} k \cdot g + w_i + o(1).$$

Remark 3. Paper [50] presents the following useful expression for the expected cost per unit time:

$$g = \sum_{j=1}^{n} \omega_j / l_{jj}. \tag{A.11}$$

Intuitively (A.11) is clear: when θ is large, the average number of visits into state E_j is θ / l_{ij}; the average cost for state E_j is equal to $(\theta / l_{ij})\omega_j$, and the total cost per unit time is equal to the sum of all these values. It was demonstrated above (see Remark 1) that for a periodic chain, (A.8) is valid. Substituting the values of l_{ij} into (A.11), we obtain the well-known relation (3.44) from Chapter 1:

$$g = \sum_{i=1}^{n} \pi_i \omega_i \Big/ \sum_{i=1}^{n} \pi_i \nu_i. \tag{A.12}$$

Formally speaking, the vector $\boldsymbol{\pi} = (\pi_1, \ldots, \pi_n)$ is not a vector of limiting probabilities; for the validity of the above formula it is sufficient that $\boldsymbol{\pi}$ is the solution of (A.10).

3. The Case when a Line Has an Arbitrary Lifetime DF

Let a line in a multiline system have an arbitrary lifetime DF $F(t)$. Consider an ordered n-dimensional sample from the population with DF $F(t)$:

$$\boldsymbol{\tau}_{(1)} \leqslant \boldsymbol{\tau}_{(2)} \leqslant \boldsymbol{\tau}_{(3)} \leqslant \cdots \leqslant \boldsymbol{\tau}_{(n)}. \tag{A.13}$$

From the basic course of probability theory it is known that the density of the ith value in an ordered sample is given by the formula:

$$f_i(t) = \frac{n!}{(i-1)!(n-i)!}[F(t)]^{i-1}[1 - F(t)]^{n-i} f(t) \tag{A.14}$$

(it is assumed that there is a derivative $f(t) = F'(t)$). Then

$$M\{\boldsymbol{\tau}_{(i)}\} = \int_0^\infty t f_i(t)\,\mathrm{d}t = \frac{n!}{(i-n)!(n-i)!}\int_0^\infty t[F(t)]^{i-1}[1-F(t)]^{n-i} f(t)\,\mathrm{d}t. \tag{A.15}$$

In terms of the multiline system considered in Section 5 of Chapter 3 the value $\boldsymbol{\tau}_{(i)}$ is the random instant of the ith failure in the system. Therefore, to compute the value $M\{w(k)\}$ one can use the formula (A.15), determining the ME of the time of appearance of the ith failure.

References

[1] Anderson, T. W., An introduction to multivariance statistical analyses (Wiley, New York, 1958).

[2] Andronov, A. M. and I. B. Gertsbakh, Optimum preventive maintenance in a certain model of accumulation of damages, Engineering Cybernetics, no. 5 (1972) pp. 832–836.

[3] Bailey, R. and B. Mahon, A proposed improved replacement policy for army vehicles, Oper. Res. Quart., vol. 26 (1975) pp. 477–494.

[4] Barlow, R. E., Applications of semi-Markov process to counter problems, Studies in Applied Probability and Management Science (Stanford University Press, Stanford, 1962).

[5] Barlow, R. E. and L. C. Hunter, Optimum preventive maintenance policies, Oper. Res., vol. 8, no. 1 (1960) pp. 90–100.

[6] Barlow, R. E., Hunter, L. C. and F. Proschan, Optimum checking procedures, J. Soc. Ind. Appl. Math., vol. 11, no. 4 (1963) pp. 1078–1095.

[7] Barlow, R. E. and A. W. Marshall, Tables of bounds for distribution with monotone hazard rate, J. Amer. Statist. Assoc., vol. 60 (1965) pp. 872–890.

[8] Barlow, R. E. and F. Proschan, Planned replacement, ch. 4, Studies in Applied Probability and Management Science (Standford University Press, Stanford, 1962).

[9] Barlow, R. E. and F. Proschan, Comparison of replacement policies and renewal theory implications, Ann. Math. Statist., vol. 35, no. 2 (1964) pp. 577–589.

[10] Barlow, R. E. and F. Proschan, Mathematical theory of reliability (Wiley, New York, 1965).

[11] Barlow, R. E. and F. Proschan, Bounds of integrals with applications to reliability problem, Ann. Math. Statist., vol. 36, no. 2 (1965) pp. 565–574.

[12] Bartholomew, D., An approximate solution of the integral equation of renewal theory, J. Roy. Statist. Soc., vol. 25, no. 2 (1963) pp. 432–441.

[13] Barzilovich, Ye. Yu., Servicing of complex systems, I, Engineering Cybernetics, no. 6 (1966) pp. 67–81.

[14] Barzilovich, Ye. Yu., Maintenance of complex technical systems, II (Survey), Engineering Cybernetics, no. 1 (1967) pp. 63–75.

[15] Barzilovich, Ye. Yu., V. A. Kashtanov and I. N. Kovalenko, On minimax criteria in reliability problems, Engineering Cybernetics, no. 6 (1971) pp. 467–477.

[16] Beichelt, F., Zuverlässigkeit und Erneuerung (VEB Verlag Technik, Berlin, 1970).

[17] Beja, A., Probability bounds in replacement policies for Markov systems, Manag. Sci., vol. 16, no. 3 (1969) pp. 253–264.

[18] Bellman, R., Dynamic programming (Princeton University Press, Princeton, N.J., 1957).

[19] Belyayev, Yu. K., B. V. Gnedenko and A. D. Solovyev, Mathematical methods of reliability theory (Academic Press, New York, 1969).

[20] Berg, M. and B. Epstein, Comparison of age, block and failure replacement policies, Oper. Res., Statistics and Economics Mimeograph Series, no. 153 (Technion - Israel Institute of Technology, 1974).

[21] Blackwell, D., Discrete dynamic programming, Ann. Math. Statist., vol. 33 (1962) pp. 719–726.

[22] Burtin, Yu. D. and B. G. Pittel, Semi-Markov decisions in a problem of optimizing a checking procedure for an unreliable queuing system, Theory Prob. Applic., vol. 17, no. 3 (1972) pp. 472–492.
[23] Cox, D. R., Renewal theory (Methuen, London, 1962).
[24] Denardo, E. V. and B. L. Fox, Multichain Markov renewal programming, SIAM J. Appl. Math., vol. 16 (1968) pp. 468–487.
[25] Derman, C., On minimax surveillance schedules, Nav. Res. Log. Quart., vol. 8 (1961) pp. 415–419.
[26] Derman, C., On sequential decisions and Markov chains, Manag. Sci., vol. 9, no. 4 (1961) pp. 16–24.
[27] Derman, C., Optimal replacement and maintenance under Markovian deterioration with probability bounds of failure, Manag. Sci., vol. 9, no. 3 (1963) pp. 478–481.
[28] Derman, C., Stable sequential control rules and Markov chains, J. Math. Anal. Applic., vol. 6, no. 2 (1963) pp. 257–265.
[29] Derman, C., Finite state Markovian decision processes (Academic Press, New York, 1970).
[30] Doetch, G., Anleitung zum praktischen Gebrauch der Laplace-Transformation (R. Oldenburg, München/Wien, 1967).
[31] Drenick, R. E., Mathematical aspect of the reliability problem, J. Soc. Ind. Appl. Math., vol. 8 (1960) pp. 125–149.
[32] Feller, W., An introduction to probability theory and its applications, vol. 1, 3rd ed. (Wiley, New York, 1968).
[33] Fisher, R. A., The statistical utilization of multiple measurements, Ann. Eugen., vol. 8 (1938) pp. 376–386.
[34] Flehinger, B. J., A general model for the reliability of systems under various preventive maintenance policies, Ann. Math. Statist., vol. 33, no. 1 (1962) pp. 137–156.
[35] Gertsbakh, I. B., Optimum rule for maintenance of a system with many states, Engineering Cybernetics, no. 5 (1964) pp. 39–45.
[36] Gertsbakh, I. B., Selection of optimal maintenance policies for groups of identical elements in automatic system, Automation and Remote Control, no. 9 (1964) pp. 1194–1200.
[37] Gertsbakh, I. B., Optimum use of reserve elements, Engineering Cybernetics, no. 5 (1966) pp. 73–78.
[38] Gertsbakh, I. B., Dynamic redundancy: Optimal control of back-up-elements actuation, Automatic Control, no. 1 (1970) pp. 26–31.
[39] Gertsbakh, I. B., Optimal control of semi-Markov processes in the presence of constraints on state probability, Cybernetics, no. 5 (1970) pp. 597–609.
[40] Gertsbakh, I. B., Sufficient optimality conditions for control limit policy in a semi-Markov process, submitted for publication in the J. Appl. Prob.
[41] Gertsbakh, I. B. and Kh.B. Kordonsky, Models of failure (Springer-Verlag, Berlin/Heidelberg/New York, 1969).
[42] Gertsbakh, I. B., M. Maksim and V. Venevcev, The optimum computation procedure for large problems using computers, Avtomatika i Vichislitelnaya Technika, no. 3 (1967) pp. 81–84 (in Russian).
[43] Gertsbakh, I. B. and G. L. Popov, Optimum choice of preventive maintenance times for a hierarchical system, Autom. Control Comp. Sci., no. 6 (1972) pp. 24–30.
[44] Gertsbakh, I. B. and M. Prilucky, An optimum procedure for emergency-preventive maintenance of a switching system, Telecomm. Radio Eng., no. 8 (1973) pp. 45–51.

[45] Hastings, C., The repair limit replacement method, Oper. Res. Quart., vol. 20, no. 3 (1969) pp. 337–350.
[46] Howard, R. A., Markov processes and dynamic programming (Technology Press and Wiley Press, New York, 1960).
[47] Howard, R. A., Research in semi-Markov decision structures, J. Oper. Res. Soc. Japan, vol. 6, no. 4 (1964).
[48] Howard, R. A., Dynamic probabilistic models, vols. 1–2 (Wiley, New York, 1971).
[49] Jardine, A. R. S., Maintenance, replacement and reliability (Pitman Publishing, London, 1973).
[50] Jewell, W. S., Markov-renewal programming, 1, 2, Oper. Res., vol. 11, no. 6 (1963) pp. 938–971.
[51] Jorgenson, D. W., J. J. McCall and R. Radner, Optimal replacement policy (North-Holland, Amsterdam, 1967).
[52] Kao, E., Optimal replacement and rules when changes of state are semi-Markovian, Oper. Res., vol. 21, no. 6 (1973) pp. 1231–1249.
[53] Klein, M., Inspection–maintenance–replacement schedules under Markovian deterioration, Manag. Sci., vol. 9, no. 1 (1962) pp. 25–32.
[54] Klusov, I. A. and A. R. Safarianc, Rotor-type automatic production lines (Mashinostroyenie, 1969, Moscow) (in Russian).
[55] Kolesar, P., Minimum cost replacement under Markovian deterioration, Manag. Sci., vol. 12, no. 9 (1966) pp. 694–706.
[56] Kordonsky, Ch. B., Applications of the probability theory in engineering (Fizmatgiz, 1963, Moscow) (in Russian).
[57] Kotchetkov, Ye. S., J. Krzhepela and M. Ulrich, Optimal statistical sampling plans for control of one type of production devices, Kybernetika (Czech.), vol. 1 (1965) pp. 421–429 (in Russian).
[58] Kozlov, B. A., Determination of reliability indices of systems with repair, Engineering Cybernetics, no. 4 (1966) pp. 87–94.
[59] Levitin, S. M., Nonlocal and noise parameters as indexes of the gradual tubes worsening, in the Collection: Reliability of Radio and Electronic Equipment (Sovietskoe Radio, Moscow, 1958) (in Russian).
[60] Mann, A. S., Linear programming and sequential decisions, Manag. Sci., vol. 7, no. 6 (1960) pp. 259–267.
[61] McCall, J. J., Maintenance policies for stochastically failing equipment: A survey, Manag. Sci., vol. 11, no. 5 (1965) pp. 493–524.
[62] Mercer, A., On wear-dependent renewal processes, J. Roy. Statist. Soc., vol. B 23 (1961) pp. 368–376.
[63] Morse, P. M., Queues, inventories and maintenance (Wiley, New York, 1958).
[64] Noonan, G. and C. Fain, Optimum preventive maintenance policies when the immediate failure detection is uncertain, Oper. Res., vol. 10, no. 3 (1962) pp. 407–410.
[65] Pestov, G. G. and L. V. Uschakova, Optimum strategies in dynamic replacement, Engineering Cybernetics, no. 5 (1971) pp. 845–848.
[66] Pittel, B. G., On asymptotic estimates in a dynamic decision-making problem, Theory Prob. Applic., vol. 14, no. 2 (1969) pp. 244–262.
[67] Pittel, B. G., A linear programming problem connected with optimal stationary control in a dynamic decision problem, Theory Prob. Applic., vol. 16, no. 4 (1971) pp. 724–728.
[68] Polovko, A. M., Fundamentals of reliability theory (Academic Press, New York, 1968).
[69] Pyke, R., Markov renewal processes: Definitions and preliminary properties, Ann. Math. Statist., vol. 33 (1961) pp. 1231–1242.

[70] Pyke, R., Markov renewal processes with finitely many states, Ann. Math. Statist., vol. 33 (1961) pp. 1243–1259.

[71] Rao, C. R., Linear statistical inference and its applications (Wiley, New York, 1965).

[72] Romanovskiĭ, I. V., Asymptotic behaviour of dynamic programming processes with a continuous set of states, Sov. Math. Dokl. 5 (1964) pp. 1684–1687.

[73] Romanovskiĭ, I. V., Existence of an optimal policy in a Markov decision process, Theory Prob. Applic., vol. 1, no. 1 (1965) pp. 122–127.

[74] Romanovskiı, I. V., The turnpike theorem for semi-Markov decision processes, Proc. Steklov Inst. Math., no. 111 (1970) pp. 249–267.

[75] Rozenblit, P. Ya., Estimating optimal preventive maintenance strategies, Engineering Cybernetics, no. 5 (1973) pp. 758–761.

[76] Schaeffer, R., Optimum age replacement policies with increasing cost factor, Technometrics, vol. 13 (1971) pp. 139–144.

[77] Shiryaev, A. N., Some new results in the theory of controlled random processes, Transactions of the Fourth Prague Conference on Information Theory, Statistical Decision Functions, Random Processes (Prague, 1967) pp. 131–203.

[78] Shiryaev, A. N. and O. V. Viskov, On control reducing to optimal stationary modes, Trudy Matem. In-ta im. V. A. Steklova, vol. 71 (1964) pp. 35–45 (in Russian).

[79] Smith, W. L., Regenerative stochastic processes, J. Roy. Soc. (London), vol. A232 (1955) pp. 6–30.

[80] Taylor, H. M., Optimal replacement under additive damage and other failure models, Nav. Res. Log. Quart., vol. 22, no. 1 (1975) pp. 1–18.

[81] Voskoboyev, V. F., Allowance for incomplete renewal in control of the state of a technological system, Engineering Cybernetics, no. 4 (1971) pp. 650–657.

[82] Weiss, G. H., A problem in equipment maintenance, Manag. Sci., vol. B, no. 3 (1962) pp. 266–277.

[83] Weiss, G. H., Optimal periodic inspections programs for randomly failing equipment, J. Res. Nat. Bur. Stand. B, vol. 67, no. 4 (1963) pp. 223–228.

[84] Wolfe, Ph. and G. B. Dantzig, Linear programming in a Markov chain, Oper. Res., vol. 10, no. 5 (1962) pp. 702–716.

[85] Zukhovitskiy, S. I. and L. I. Avdeyeva, Linear and convex programming (W. B. Saunders, Philadelphia, 1966).

INDEX